1

Into the

Doorways

P.S. Winn

© 2016

*

*

*

Chapter 1

Taking a step, Jess Holdens' foot slipped on the black ice and he went down on one knee. Cussing under his breath, Jess was able to straighten his leg and stand. Looking up and down the empty highway, he shook his head wearily. The last ride he'd gotten had dropped him off five miles back. After walking from the small city he thought had been called Ransburg, Jess had only seen three cars and none had even slowed down at the sight of his thumb sticking out in the cold morning air.

Jess adjusted his black leather jacket. The damn thing looked good, but didn't work well at all when it came to keeping out the cold, and it was more than just cold, it was freezing out here. The last few days hadn't been all that bad. Then a northern cold front had blew into the area. Now, just a month from spring, the weather felt like the dead of winter. All Jess wanted right now was a warm spot to lay up.

Actually, the place he wanted, no, that he needed to be, was a ways up the road. Not only did Jess want out of the cold, but he knew he was running out of time to reach his destination. He knew Holston was still a long way up the road and that was if he was in someone's nice warm vehicle. Hell, why had he bet Kyle that he could make it two weeks in this place. No one he knew had ever tried to make it more than a week.

Clenching his fingers, Jess felt the motion painful and knew that feeling was from more than the cold weather. Shaking his head at the thought, Jess looked down the road and felt his spirits lift at the sight of an older black car headed toward him. Jess wasn't any kind of authority on cars, but he was willing to bet the one barreling toward him was at least fifty years old.

As the car rushed past him on the snow covered road, Jess felt like screaming. As he saw the brake lights glow red and the back end of the car fishtail with the motion of the driver braking, Jess grinned. Before the driver could change his mind, Jess ran as fast as he could to the passenger side door.

A man in his late twenties already had the window rolled down when Jess got there. "Where you headed?"

Pushing back his shoulder length, brown hair, and then pointing up the road, Jess spoke, out of breath, from his short run. "Just up to the next town."

The driver, also in his twenties, leaned forward. "You mean Holston?"

As soon as Jess nodded, the driver smiled. "Well, hop on in, we're going that way. You look like you could use a ride and a rest."

Rubbing his cold hands together, Jess smiled. "I really appreciate it."

Jess waited for the guy in the passenger seat to open the door and then lean his seat forward so he could crawl in the back. As he sat down, Jess looked at the various fast food wrappers scattered on the floor. He also saw quite a few empty beer cans scattered amongst the garbage. Knowing he was lucky to get the ride and knowing his time was running out, Jess ignored the mess. "I sure am glad you stopped. This is really a great car, what year is it?"

Patting the steering wheel, the driver smiled broadly. "She's a work in progress. A sixty two Chevy Impala SS. She must have been made about the time you were born."

A frown creased Jess' forehead, he wasn't much older than the two men who occupied the front seats. It took all his self-control to stop himself from leaning over the seat to get a look at himself in the rear view mirror. Taking a breath, Jess collected himself. "I guess I must look pretty bad. I've been catching rides the last few days trying to get to Holston. Once I get there, I'll be okay."

The guy in the passenger seat turned so he could look at Jess. His brown eyes squinted beneath the red hair that slid down over one eye. "Don't you got a car dude?"

Jess nodded, he'd been asked this question before and had come up with what he hoped was a plausible answer. "It's in the shop, damn thing's pretty totaled though. Probably be cheaper to buy a new one."

The driver was nodding. "Ain't that the truth? Hell, those shops charge you an arm and a leg for the simplest repair. I'm glad I can do most of the work on this car myself. By the way, I'm Mick and the guy over there is Danny."

A grin crossed Jess' face, glad the two hadn't asked more about the wrecked car he didn't really own. "I'm Jess, thanks again for picking me up."

Eyes still focused on the road, Mick nodded. "No problem, just sit back and relax. We'll have you in Holston in no time."

As Jess leaned back in his seat, Mick reached down and turned up the radio. Jess didn't know the song, but by the way Mick was beating out the rhythm out on the steering wheel, he did.
Looking down at his hands, Jess was surprised and more than a little nervous at the sight. The skin, stretched tight, was covered in age spots. Flexing his fingers,

8

Jess felt the ache of what he knew had to be some type of arthritis. Looking toward the dashboard up front. Jess could see the speedometer sitting at eighty five. Despite the ice covered roads, Jess almost wished they were going even faster. He knew, for him, time was running out.

Turning to look out the window, Jess didn't see the landscape. Instead he saw an image of Bailey Conners. He knew she'd be worried and probably damn mad. She hadn't wanted him to come here. Jess wished now he would have listened to her arguments against his journey that had begun with the bet between him and Kyle. He should have known two weeks out here was too long.

Jess had met Bailey five years ago when both of them had gone to work for the Postern Project. Kyle and Patrick had been added to their team two years after that. Their team had originally been made up of five members. They'd lost Marna a year ago in a tragic death. She'd never been replaced, so now it was just the four of them working under Joshua Porter.

Jess and Bailey had went from coworkers to

partners, but both Kyle and Patrick remained bachelors. Jess didn't blame them, given the nature of their work.

In the front seat, Dannys' voice interrupted Jess' thoughts. He was pointing to a road sign as the car rushed past it. "Holton, five miles, looks like we're almost there."

Relief flooded through Jess, maybe he would get there before it was too late. He leaned forward in his seat. "I just need to be dropped off at the City Limits' Sign. I have some friend who are going to be meeting me there."

Taking his attention off the road, Mick frowned and then glanced back at Jess. He was going to ask why in the hell Jess' friends hadn't just picked him up, instead of making him hitch in the cold, but the sight of the man in the backseat froze the words. Less than an hour ago, he'd picked up what he thought was a fifty year old hitchhiker. Now, the man occupying the back seat was at least twice that age. Finally Mick found his voice. "What the hell…"
As he spoke, Micks' hand gripped the

steering wheel and pulled to the right as he turned further in his seat to get a better look at his strange passenger. The car lost traction and started to slide. Instead of hitting the brakes, Mick tried to turn the wheel to the left and then back to the right.

In the passenger seat, Dannys' hand went up to cover his face as the car flipped over and skidded off the road and into the shallow ditch that ran alongside the highway.

All three men screamed as they were tossed upside down inside the car. As the car came to a stop on its' roof, Jess realized his screams were the only ones still echoing in the cars' interior.

Finding himself lying on what have should have been the floor of the car, but was now the roof, Jess rolled over, moaning with pain.

He could see Mick and Danny laying in front of him, both covered in blood. Reaching a shaky hand forward, Jess touched Dannys' neck first. By the angle of the young man's head, Jess wasn't surprised at the lack of a pulse. Turning to Mick, Jess didn't bother checking for a pulse. A piece of shattered glass from the windshield had embedded itself in the drivers' neck, almost slicing the head off at Micks' shoulders.

Letting out a strangled curse, Jess turned away from the horrendous sight. Looking around, he saw all the back windows had miraculously and unfortunately stayed intact. He needed to get out of the car. Not only did he need to get to the city limits' sign and his friends, but he definitely didn't need the police to come by and find him here in the wreck. The last thing he wanted was to try and explain his presence here.

Dispensing with the idea of trying to climb over the two dead bodies to reach the front of the car, Jess tried lifting both legs in an attempt to kick out the backseat windows. Instead, only one leg moved, the other only sent a shocking pain through his body as Jess screamed in agony.

Holding his bad leg with one hand, Jess gritted his teeth and kicked the window over

and over with the other leg. Finally, he heard the sound of the glass breaking.
Maneuvering around, Jess reached over and pulled himself through the window. As his hands touched the ice covered ground, Jess felt another pain in his already useless leg. Reaching forward, Jess found a rock to grab onto. Using it, he pulled himself forward. Another scream echoed in the cold air as a piece of glass dug deeper into Jess' leg, leaving a four inch gap.

Lying in the ditch, free now from the car, steam poured from Jess' mouth as each jagged breath hit the cold air.

It took a moment for Jess to recover a little and steady his breathing. Rolling over onto his back, Jess turned his head and looked down at his leg. He watched as blood poured out through the rip in his jeans. Reaching down, Jess clasped his hand over the tear and pushed down as hard as he could. Blood still flowed out and onto his hand, warm at first, it cooled quickly in the frigid air.

Still holding his leg, Jess turned back onto his stomach. Looking up the road, he saw the Holston City Limits' sign. Although only about thirty feet ahead of him, it might as well be thirty miles.

Drawing a deep breath, Jess half crawled across the ground. Using his one good leg and dragging the other behind, Jess inched forward toward the sign. Sweat broke out on his forehead and then froze in place. Wiping it away, Jess left behind jagged scratch marks that oozed blood.

Halfway to his destination, Jess felt his vision begin to fade. Grabbing his leg, Jess clamped down as hard as he could, hoping to drive away the darkness.

Instead the shadows took over and Jess fell flat on the ground as he passed out.

Chapter 2

In one of the rooms at the Postern Projects'
facility, Kyle Wright and Patrick Baker sat
at a table having coffee. They looked up
when the door to the room slammed open.
The brown haired, blue eyed woman who
stood a few inches over five foot, glared at
the men as she slammed the door shut
behind her. "What the hell are you two
doing? Why aren't you over there checking
the viewer?"

Stepping over to the table, Bailey slammed a
hand down on the top. "Damn it, Jess is
overdue. What the hell is the matter with
you two?"

Both sets of brown eyes stared at the
woman. The color of their eyes was the only
similarity between the two men. Brown
haired Kyle, was short and stocky, while

blonde haired Patrick, his long hair pulled back in a ponytail, was tall and slender.

It was Patrick who shook his head. "We've been checking the viewer every hour Bailey. No sign of him yet. If we check more often than that someone will get a red flag. You know how Porter feels about any of us traveling without permission."

Folding her arms across her chest, Bailey shook her head. "I don't give a damn what Joshua Porter thinks. Jess is in trouble." She turned to face Kyle. "Why did you make that bet with him? You knew Jess would never be able to resist a challenge. Damn it Kyle."

Turning back to Patrick, Bailey frowned. "When's the last time you checked?"

Looking at his watch, Patrick shrugged. "About a half an hour ago, listen Bailey, we're both worried about him too."

Looking over at a computer station in the corner, Bailey pointed. "Open up the viewer, I want to see for myself."

Patrick stared at Bailey. "We really should wait a little while. We have a meeting with Porter in a couple hours. If he knows we've been traveling, we're in a world of trouble."

The blue eyes flared with anger. "Open the viewer Patrick, now. I have to take a look, that's Jess out there."

Reluctantly, Patrick stood and walked over to the computer. Bailey stepped up behind him, followed by Kyle a few seconds later. All three watched the screen as Patrick pushed the buttons to activate the viewer position to focus on the area by the Holston City Limits' sign. That was the place they all had agreed on Jess being when it was time to return.

As the picture on the screen flickered to life, Bailey let out a short scream. "Oh hell, look at that."

The three people stared at the car lying on its' top. Bailey pointed at a lump on the ground not far from the car. "That's Jess, oh my God, that's Jess. Look at his jacket."

17

All three recognized the leather jacket that Jess always wore. Turning to Patrick, Kyle grabbed his arm. "Open that door Patrick. We have to help him, we have to get Jess."

Knowing Bailey was right, Patrick still hesitated. Their whole group was going to be in a lot of trouble from Joshua Porter for unauthorized travel. Still, they couldn't just sit here with Jess lying out there.
Patrick turned to the console and pushed a few more buttons before turning to Bailey and Kyle. "I'll have to stay here in order to keep the door open long enough. Can you two get Jess and get back here without me?"

Baileys' head bobbed up and down. "I'll get him back here myself if I have to."

Kyle shook his head. "You don't have to worry about that Bailey. I'm sorry we didn't check that viewer more often."
Kyle grabbed Baileys' hand. "C'mon, let's go get Jess and bring him home."

The two walked over to what looked like a blank wall. Just as they got there, Patrick pushed the controls and a shimmering

doorway appeared. Looking through it was like looking through a heavy rain shower, the image outside distorted.
Kyle and Bailey stepped through together and tensed as the freezing air hit them. Ignoring the cold, Bailey ran to where Jess lay on the side of the road. Behind her, Kyle looked up and down for any witnesses to their arrival in this world before he too ran over to Jess.

Baileys' eyes widened in shock at the sight of Jess' head of thinning white hair. When he'd left the Postern Project building, he'd entered this world with shoulder length thick brown hair that was just a shade darker than Baileys'. Now that hair, what was left of it, was as white as the snow on the ground.

Kneeling down, Bailey touched Jess' cold face, then felt his neck. She turned back to Kyle. "He's alive, but just barely. His pulse is faint and erratic. We need to get him back inside the project."

Knowing that moving Jess might make his injuries worse, Kyle also knew that not moving him from this world would be like signing a death sentence for his friend. Leaning down, Kyle took a hold of Jess under his arm pits and pulled him toward the doorway.

Knowing she'd be more hindrance than help, Bailey walked beside Kyle, scanning Jess' body as they went, trying to take note of his numerous injuries. Just before they got to the doorway, Bailey looked at Kyle. "I need to go and check that car, someone might need our help."

Out of breath, Kyle only nodded. Bailey ran quickly to the car. Bending down, she stared inside. With one look, she knew it was too late to help the occupants. Bailey sadly turned away. Closing her eyes, she drew several deep breaths and then wrapping her hands around her freezing body, ran to where Kyle was just pulling Jess through the doorway.

As the three made it through, Patrick closed the portal doorway behind them and then stepped over to his friends. "We need to get Jess into the healing chamber."

Kyle nodded. "Can you get his legs Patrick? I don't think I can take him alone."

Patrick went to grab Jess' legs, as he grabbed them, he turned to Bailey. "Go get the chamber ready, we'll be right there."

Nodding, Bailey ran out of the room, leaving the door open behind her for the others. Running down the hall, she passed two project workers who stared at her with frowns covering their faces. Ignoring the looks, Bailey went quickly to the end of the hallway and opened the door to the room that held two of the special healing chambers. She let out a sigh of relief, seeing neither was being used.

Stepping over, she opened one of the chambers, hating that it looked more like a funeral casket than a healing chamber.

By the time Bailey had finished setting up the unit, Kyle and Patrick were carrying a still unconscious Jess into the room.

After the two laid their companion in the chamber, Bailey reached down and brushed the now almost pure white hair off his forehead before touching his face with a trembling hand. She barely recognized her work and life partner. Jess' face was creased with age. The twenty seven year old looked now more like a century year old man. Bending over, Bailey kissed the face she loved before closing Jess in the chamber.

As soon as the lid closed, Patrick was busy at the control panel adjusting the dials and then setting the healing process into motion. Bailey had pulled over a chair and was now sitting next to the chamber, wringing her hands and staring at the container that held Jess, the man who she lived with and loved for three years after spending two as his assistant only.

Kyle touched Bailey gently on her shoulder. "Maybe we should go and grab a cup of coffee."

Patrick turned from the control panel and nodded. "Kyle's right Bailey, you can't do anything more for Jess. It will be an hour before we know if we got to him in time."

Patrick regretted the words the moment he said them, but Bailey just shook her head.

"I'll wait here. You two go on. I'll be okay."

The two men exchanged glances before Kyle nodded. "We'll bring you back a cup. Don't forget we also have that meeting today with Joshua Porter."

Pulling her eyes reluctantly from the chamber. Bailey looked at Kyle. "We should know before then shouldn't we? I mean that Jess is going to be okay, right?"

Looking at his watch, Kyle nodded. "It'll be close, but yeah, we should know."

With one more glance at Bailey, whose attention had reverted back to the container, Kyle and Patrick left the room.
They closed the door as they left, giving Bailey her privacy.

As soon as the door closed behind the men, Bailey allowed herself to break down and the tears cascaded down her cheeks. She wanted to be mad at Kyle for making the bet and at Jess for taking it, even at Patrick for being afraid to check the viewer sooner. All she felt right now though, was the worry for Jess' health. She couldn't help the doubts that crossed her mind. Had they got to Jess in time? Could the healing chamber fix all the damage, not just the aging, but the injuries from that awful car wreck too? Bailey sighed, knowing the next hour would crawl by at a snails' pace. Reaching out, she rubbed the top of the chamber and whispered. "Hang in there Jess, don't leave me, please don't leave me."

Pressing her head against the chamber, Bailey thought back to when she had first seen Jess. She had just been asked to join the Postern Project and it had been Joshua Porter who had escorted her to the lab where Jess was sitting in front of a computer screen. He hadn't even heard them come in the room until Joshua had yelled. "Jess, are you with us?"

Turning, Jess had laughed, his brown eyes shining with humor. Pushing the brown hair that hung to his shoulders back, Jess shook his head. "Sorry, I didn't hear you come in."

Jess looked at Joshua first, but then he took the time to study the woman who stood next to his boss. Bailey could tell Jess liked what he saw, but she wasn't here for her looks. She had a damn fine brain behind the blue eyes and the slightly turned up nose that she hated. Joshua placed an arm around Baileys' shoulders. "Jess, I'd like you to meet Bailey, she's going to be joining you. Now that you have a partner, the two of you are the first two members of the Alpha Team. I need you to show her what we've been doing. She's been filled in on the traveling and gone through the simulators. Now, it's time for her to take a real journey and see what all this is really about."

Jess stood and took Baileys' hand, holding on to it while he spoke. "It's good to meet you Bailey. If Joshua brought you here, you must be as smart as you are pretty. Welcome to the Alpha team."

Pulling her hand out of the grip even though she liked the feel of Jess' strong hand, Bailey grimaced. "I don't mind being smart, but while I appreciate the other compliment,

I don't think it will do me much good in this job. I'm glad to be part of the team though."

From the beginning, Bailey and Jess had a special connection. It took two years though before the two decided that it would be okay to give romance a try without hurting their working relationship. To their delight, and relief, being romantically involved only enhanced the work they did for The Postern Project. The two had been together professionally and personally ever since.

In the healing room, rubbing the top of the container, Bailey tried to smile. "You can do this Jess."

In the projects' cafeteria, Kyle and Patrick got their coffees and sat down at one of the tables. The other tables in the cafeteria were for the most part, empty. It was between the lunchtime rush and the evening crowd. Kyle absently turned the coffee cup in front of him around and around. Finally he realized he was doing it and stopped, then looked up at Patrick, rubbing the perpetual stubble on his chin. "What if we were too late? Damn, I wished we would have checked the viewer sooner."

Patrick shook his head. "We looked a half an hour before Bailey came in Kyle. I'm sure Jess is going to be okay. You should be more worried about that meeting with Porter. If he hasn't found out about Jess' unauthorized trip already, he's gonna want to know where Jess is and why he's not at that meeting. All the teams are supposed to be there. In an hour we'll know if the chamber is helping, but he'll still have to stay in that thing at least another six hours for full recovery." Kyle sighed. "We better tell Porter the truth. He'll find out sooner or later. It might sound better if we just come clean with the whole damn story."

Kyle stood up. "Let's get Baileys' coffee and get it back to her. I'd rather be there waiting than here."

Nodding, Patrick also stood. "Yeah, me too."

By the time they got the coffee and returned to the healing room, they only had to wait a half an hour before they decided enough time had passed for them to check Jess' progress.

All three held their breaths as they lifted the lid and then stared in at Jess. Smiling at the sleeping man in the chamber, Bailey then turned to Kyle and Patrick, her blue eyes sparkling. "It's working, look at his hair."

Both men nodded, it was easy to see the difference in the hair, white when they had put Jess in the chamber, it was now more of a salt and pepper color. Reaching down, Patrick ripped Jess' pant leg where the glass had already begun a tear. He smiled broadly when he saw the almost healed wound. "I think we got to him in time."
Patrick looked over at Bailey. "I'm sorry we didn't check that viewer sooner Bailey."

Kyle nodded. "Me too, we shouldn't have been worried about what Porter would do. Jess is more important to us than anything else."

Bailey smiled. "He's going to be okay, that's all that matters. Leaning down, Bailey kissed Jess' cheek. "You just have a good rest Jess, we'll take care of everything else."

Carefully closing the lid, the three stepped back, staring at the box. Patrick was the first to step away as he went over to the console and restarted the healing chamber. Still staring at the top of the chamber, Bailey sighed. "I wish he'd been awake."

Kyle smiled. "The more sleep he gets, the faster he'll heal. I think the three of us better get back to the lab. We need to get our reports together for that meeting. I don't know if Porter will ask to see them, but I want to have them ready, just in case. Besides, Jess will have to be in the chamber for a while, it won't do any of us any good to be here."

Blowing a kiss toward the container, Bailey then turned and followed Kyle and Patrick to the lab. The three got to work, gathering any information from their last few travels that they thought might be needed in the meeting. When they'd gotten everything assembled, Bailey, turned to face Kyle and Patrick frowning, her eyes worried. "What about Jess' last trip? Are we telling Joshua Porter about it?"

Kyle looked at Patrick before turning back to Bailey and nodding. "I think we better. Jess did stay in that world longer than anyone else. Maybe we can learn something from that."

Bailey sighed. "A few days longer and he's four times older than when he went in. I don't think that will solve any problems we're looking at."

It was Patrick who answered. "Terrene needs all the help it can get. Every bit of information gets us a step closer to a new home."

Nodding, Bailey shrugged. "I guess it does and we're running out of time."

Baileys' last words echoed in their minds as the three left the lab and headed to Joshua Porter's conference room. When they stepped in, they saw the other teams were already seated in the room. Teams Beta and Gamma had five members each. With only three of their already smaller team present, the Alpha team knew they'd really stand out. Besides, because they were Joshua Porter's Alpha team, they had more experience and were supposed to be the ones the others looked up to. All of them knew that fact would only make things worse for them.

Joshua wasn't anywhere in the room yet, so Bailey, Kyle and Patrick took seats closest to the back of the room. Although their nerves kept them silent, they could hear the others talking quietly amongst themselves. Mostly wondering what this meeting Joshua Porter had called was all about.

The murmuring ceased as a door opposite the one the Alpha team had come through opened and a man in his forties stepped in the room.

Joshua Porter had sandy hair and light green eyes. He was a few inches over six foot and had a muscular build. Of course everyone in the Postern Project had to stay in good physical condition.

Joshua grinned at those in the room, and the eyes lit up. "So glad everyone could make it today."

As he said those words, Joshuas' gaze fell on the three in the back of the room and Bailey thought she saw something in the green eyes close to merriment. She was certain then that Joshua Porter knew all about Jess' journey to the Parallel World called Earth.

Walking to the podium in front of the room, Joshua cleared his throat. He looked down at the podium and then reaching down shuffled through some papers that laid there. Grabbing one, he lifted it and held it in front of him. When he looked up, his eyes were just the opposite from what they had been a moment ago. Now the green was a shade darker and Bailey looking at Joshua Porter, knew that no one in the room was going to like whatever news it was that their boss had to share.

Joshua had dreaded this moment for years. That's why he had started the Postern Project all those years ago. Two years before he had put together the first Alpha team, Joshua knew what was coming.

When a group of scientists had first gotten a glimpse of the Asteroid Hartman, named after the main scientist who had made the discovery, Professor Aaron Hartman, they had come to Joshua. At the time, no one had valid proof that Asteroid Hartman would actually hit Terrene. All the projected trajectories though had pointed to the fact. As the years went by, an inevitable hit was guaranteed. Now, they had what was more than just an approximate date.

Joshua cleared his throat again. "We've been given every possible scenario our computers could come up with concerning Asteroid Hartman. Unfortunately, as we all know, our world is in the direct path of the asteroid. Now we know that an impact that will obliterate our planet will happen in six months' time."

Joshua shook his head. "We have tried several preventative measures with no success. Our worlds' only hope now is the Postern Project. We have to find a portal to an alternate world that will sustain life. We are going to have to escalate our program. A lot of people are going to be depending on one of our teams to find that new world for them. I hate to bring you this news. I know your teams have been searching and although we've been unsuccessful, I hold a great hope that we can find a place for our people to not only survive, but to flourish."

Joshua took a breath as he scanned the small crowd. "If you'll head back to your labs, you'll find your groups' itinerary has been loaded into your computers. You'll also find new guide books and further information on what changes have been made. I don't have to tell you how important you are to Project Postern and to this world. I want to personally thank each of you for your hard work so far."

As the people from the groups began to rise, Joshua looked at the three who had just stood up from their places in the back of the room. "Alpha team, I'll need you to stay behind a moment. There's a few things we need to discuss."

Looking at Joshua, the three knew this was a directive they couldn't dismiss. All three also had the distinct feeling they were in trouble. They sat back down and waited.

Seeing the looks of apprehension crossed with that of guilt on their faces, Joshua was reminded of himself when he was much younger.

He'd gotten in a lot of trouble, especially for a year after he turned seventeen. The year his dad died. Joshuas' dad, Steve Porter, had been a single parent. That job wasn't one Steve had wanted. Joshuas' mom, Tina Porter, had died giving birth to Joshua. Tina had been in her late forties when she'd lost her life bringing her only child into the world. Joshuas' dad had been in his early fifties and only lived to be sixty nine.

It was Joshuas' science teacher who finally opened his home to the teen who had suddenly became a troublemaker. Before that, Joshua had been a straight A student. Paul Garrison knew Joshua had potential. The science teacher took Joshua under his wing and by his high school graduation,

Joshua was at the top of his class.

Paul Garrison was more than a high school teacher, he was also a fairly well known and well respected scientist.

It was in his garage, tuned lab, where Joshua was first introduced to the possibility of Parallel Worlds.

At the time, while others were talking about traveling to other planets, Paul had a different journey in mind. Thankfully for Joshua, Paul took him along for that ride. Joshuas' teacher, friend, mentor and really his only family had passed away two years back. Paul had died after a courageous battle with lung cancer. The doctors blamed Pauls' heavy smoking, to Joshua though, the worry was always in his mind that Paul had gotten that cancer on one of their experimental jaunts to Parallel Worlds.

Realizing the three members of the Alpha team were staring at him, Joshua pushed the memories of Paul to the back of his mind. He'd bring them out to examine another, more appropriate, time.

Looking from one face to the next, Joshua finally let his gaze settle on Bailey. "How's Jess doing after that journey of his?"

Trying not to show her surprise, Bailey shrugged. "I think he's going to be okay."

Joshua turned one of the unoccupied chairs around and sat so he was facing the team. "That was a damn fool thing to try."

Knowing they couldn't deny Jess' trip, Kyle shook his head. "Bailey was dead set against it. If you need someone to blame, then I'm the guy you should be looking at."

Leaning forward, Patrick shook his head. "No way, we're a team and we take the blame as one. Jess had made up his mind to go, but we could have stopped him. Maybe we were all hoping he could make it this time."

Staring first at Patrick, Joshua then turned to look at Kyle and Bailey. As his eyes settled on the only female in the Alpha team, he noticed her blue eyes narrowing.

"How long have you known?"

A half grin curved about Joshuas' lips. "Almost from the first moment you opened the doorway."

Bailey frowned. "Why the hell didn't you just stop us?"

Joshua shrugged. "I guess everyone here was hoping you could find a way to make it in the Earth World. We're running out of time and haven't had much luck so far. That doesn't mean your team is in the clear. I'm sure all of you are well aware of the no unauthorized traveling law. It's there for a good reason. We can't have teams heading out all over place. For now, any punishment will have to wait. Just know that your team will be more closely monitored than it was before. Like the other teams, you are going to be asked to continue traveling and at an accelerated rate."

Joshua shook his head wearily. "We might have six months until Asteroid Hartman hits Terrene, but we need a place well before that time. We're not as populated as some of the worlds we've found, but when the time comes, we're still going to have to move over a hundred thousand people through a doorway. Before we even announce this newest information about the date of impact, I want to have a place for the people to go. We're already planning on a worldwide panic. Hopefully knowing there's a new place set up to go when that cataclysmic event hits will help relieve some of that panic in the people."

Joshua laughed harshly. "Even then, nothing is going to be easy."

Listening to Joshua, Kyle could picture the reaction and that's if they really were able to find a world for them to travel to. "How many worlds have you set up for us to look at?"

A sigh escaped from Joshua. "Right now we've found nine that are registering as compatible. For now, each team has been assigned three of those worlds while we are busy looking for more."

Now it was Patrick whose brown eyes narrowed skeptically. "We thought the Earth World was well-suited. I mean their atmosphere almost matched our own. What happened there Professor Porter? Why did the team age so quickly?"

Shaking his head, Joshua lifted his shoulders in a shrug. "The Earth World is causing their own destruction. To think of what they've done to their World is an abomination. That world has been polluted, the inhabitants are always at war and it doesn't look like either of those things will be fixed. We have copies of their history on our computers now and it seems that place has been doing the same things over and over and never learning from their mistakes. They may have more time than we do, but believe me the world called Earth is doomed to perish."

Patrick looked at the professor, curiosity in his brown eyes. He brushed back his long blonde hair that had come out of the usual ponytail. "But, those people didn't age like we did when we went there. Why would that be?"

Bailey nodded. "I wondered that too. Jess was twenty seven when he went back through the doorway, but he was closer to a hundred less than two weeks later. That doesn't make sense."

Nodding, Joshua then shook his head. "Maybe if we would have entered that world fifty or a hundred years ago we could have found a way to survive. A persons' body can change and adapt, but that takes time. Unfortunately that's something we don't have much of."

Joshua turned to Bailey. "I've asked the doctors to monitor Jess in the healing chamber. So far, everything seems to be going well, but until he can be removed, we won't know what his final prognosis is. He's damn lucky you all got him back through the doorway when you did."

Josh pointed at Kyle and Patrick. "I want you two taking Bailey and heading back to your lab for now. Jess is in good hands and there isn't thing you could do anyway, except wait and worry. I think preparing for your mission will be a much better use of your teams' time. By the time Jess has finished his healing time and whatever rehabilitation he's going to need, you can fill him in and prepare him for traveling with you. In spite of what your team has done, the four of you are one hell of a team. I've decided, for now, the best thing is to set aside your violations and move ahead. Now…"

Joshua stood. "Get out of here and get busy. I've done more explaining and talking today than I have in the last year. Don't worry about Jess. You can visit him later tonight. I'll be checking in on him personally. I want the three of your to do something productive in the time you're waiting to go see him."

The team members, knowing they'd been given a big reprieve, didn't waste any time getting out of the conference room and back to their lab.

Chapter 3

When the three were back to the safety of their lab, Patrick went straight to the computer to see what new files and information had been added.

Bailey still standing, picked up the folders that had been left on one of the desks and absently thumbed through them.

Kyle sat down and ran a hand through his dark hair. "I hope you two realize how lucky we were not to catch hell from Porter."

Baileys' eyes opened wide. "Are you kidding me? The only reason Joshua didn't throw the book at us was because of his worries over the date that asteroid is going to come crashing down. I wouldn't call that lucky."

Giving Bailey his best, please forgive me, schoolboy grin, Kyle held up his hands in surrender. "You know I didn't mean getting hit by an asteroid was lucky. I meant...oh hell...you know what I meant."

Bailey smiled. "Okay Kyle, you're off the hook and forgiven, this time anyway. Now, let's go see what's got Patrick so mesmerized over there."

Pushing two chairs over, the two sat down, one on either side of Patrick.
Looking at Bailey first, then over at Kyle, Patrick laughed. "Glad you two decided to join me. Would you like to take a look at Province?"

Bailey frowned. "What in the world is Province?"

Kyle pointed at the computer screen. "Better rephrase that Bailey. You should have asked what world is Province."

Looking at the screen, Bailey could feel the excitement she always did when they were given a new world to explore. She turned to

Patrick. "Okay, what have you learned about this place so far?"

Clearing his throat, Patrick sat up straighter in his chair, lowering his voice to a steady tone so he sounded like a documentary announcer. "Province is a parallel world, like all the others. I'd say it's approximately three times the size of Terrene. That is in area, not population. Actually Province has about the same population as our world, maybe a little less."

Patrick glanced at his two companions. "Which means there is room for growth, that's one plus. The atmosphere is similar to ours also."

Kyle smiled. "Plus number two."

Patrick nodded. "The weather doesn't fluctuate much, even in winter we would be looking at much better weather than Terrene. That is unless you want to snow ski or partake in other winter sports."

Bailey laughed. "Or have a snowball fight. To me, the weather makes this Province a definite plus three. Now, what are the downsides?"

Patrick shrugged. "I'm afraid we'll have to find that out when we travel there. We don't have any information yet on their laws, their government or their lack of. The only way to find out the nitty gritty is to go there and do some research."

Nodding, Baileys' eyes were still dejected. "I wish there was a better way. It's hard to step into a world and know hardly anything about the people or their culture. Do you have an audio feed yet? We don't even know if they speak English in this place. You know how awful I am at trying to speak foreign languages."

Bailey turned at the sound of Kyles' laughter. "That's an understatement if I ever heard one. Have you mastered any foreign languages yet Bailey?"

The head of dark hair shook as Bailey turned to stare at Kyle. "No, damn it Kyle, you know I haven't. I just can't seem to catch on. Besides, the people in these Parallel

Worlds could speak a language none of us has even heard of before."

Facing Kyle, Patrick nodded. "She's right about that. We've been lucky so far, or maybe Joshua set that into the computers as part of his search criteria."
Turning to Bailey, Patrick smiled. "And to get back to your question, the answer is no. I haven't heard any audio. We have a few videos and some stills. It looks like all of this was done in a hurry. All of the still pictures are dated either today or yesterday. Usually they go back a couple of weeks at least. I guess Joshua is really trying to explore as many places as he can and as fast as possible."

Sitting back in her chair, Bailey sighed. "Six months, I just can't believe it. I heard Joshua say it, but I guess I haven't had the time to think about it. That's so little time. We don't have a place to go yet and when we do find one, we have to move our whole world."

Kyle looked at his two friends, his mouth set in a grim line. "That's if we find a place that's not only compatible, but one that's willing to let a whole lot of strangers in."

Patrick shrugged, but then smiled. "Maybe we'll get lucky and find an empty world just waiting for us."

Kyle shook his head, wondering how hard it would be to build a world from scratch. People were used to having homes and the technology they needed to run their lives. How many could actually survive without the things they took for granted. He shook away those thoughts, first they needed to worry about Province.

Pointing at the computer screen, Kyle frowned. "One thing I did notice already, the people we're seeing in Province dress a bit different than us."

Looking at the still shots on the screen, Bailey nodded. "We'll have to find similar styles. The last thing we want to do is stand out in a crowd."

As he looked at the pictures, Patrick smiled and pointed at the screen. "You know what? I've seen these clothes before. It was when

we were in the Earth World. Don't you remember? We were in the library."
He turned to Bailey. "Don't you recall how strange it was? Everyone was dressed funny and then we realized it was some kind of throwback day at the place. What did they call that?"

Nodding, Bailey smiled. "I do remember, that happened before we realized we were all aging so rapidly."
Putting her finger to her lip, Bailey frowned, thinking. "I got it, they called it a sock hop. I thought it was a strange name because everyone still had their shoes on."

Patrick turned back to the computer. "We should have Earth's historical records in our files. Maybe we can figure out a way to make the clothes we have look like we've just been at a sock hop. Then we can fit in with the people of Province."

The three spent the next hour going through what information they had on Earths' past history and printing out several pictures of the fifties era clothing that the people of Earth had once worn.

Bailey picked up the pile of papers and slowly leafed through them. "I think we just need to make a few adjustments to what we have in our wardrobes, then we should be able to fit in perfectly."

Kyle, not interested in clothing, noticed something else about Province. "Look at the cars, they're all driving the same model of vehicle."

Patrick frowned. "I hadn't noticed that before, but you're right. They're different colors, but other than that the cars are identical."

Bailey frowned. "How many viewers are set up out there?"

Turning to the computer, Patrick tapped the keys for a minute, then shrugged. "From what I can tell, there's only the one viewer out there. Looks like with the new expedited journeys were going to do less preliminary searches before our teams get sent out."

Bailey frowned, not liking this new development, but understanding why it had to be this way. "I think that six month deadline is going to be changing a lot of things."

Kyle stood. "If we don't do our part and find a world for us all to go to, the changes will be a hell of a lot worse."

Rubbing her eyes, Bailey nodded. "I just hope Joshua will let Jess join us." Thinking of that, Bailey looked at her watch. "No wonder I'm tired, we've been in here almost six hours. Let's shut down and go check on Jess. He should be out of the chamber by now."

Both Kyle and Patrick were more than happy to call it a day. They were both also concerned about their team mate who also happened to be one of their best friends. In fact the small four member team were together as much when off work as while they were working. With Baileys' help, the two men closed down the lab before all of them headed for the healing room.

When they stepped in the room, they were surprised to see the lids standing open on both of the healing chambers. It was easy to see both chambers were also empty.

Bailey frowned. "They must have moved Jess to a recovery room."

She stared at Kyle and Patrick. "That's a good sign, right?"

Although neither of the men knew the answer, they both quickly nodded, hoping to quell Baileys' anxiety. Kyle put an arm around Baileys' shoulder. "Let's go down and check the infirmary. I'm sure that's where we'll find him."

Patrick laughed. "And probably driving everyone nuts trying to get out of the place."

Making their way through the facility, Bailey hadn't noticed before how big the place was. It seemed to take an eternity before the trio found themselves at the nurses' station just outside the infirmary. The woman behind the desk looked up. "Can I help you?"

Unsure of how much the nurses and doctors were told about Jess' injuries, Bailey nodded. "Our friend Jess Holden was, um, he was, sick and we were just checking to see if he was brought in here."

The nurse nodded, but looked down at a paper on her desk before looking back up. "Are you part of the Alpha team?"

Kyle laughed. "We're not just part of it, we're the whole damn team."

Giving Kyle a glare, she turned to Bailey. "Mr. Porter left authorization for Alpha team members to visit. You'll find your friend in room three."
Pointing at a hallway, the nurse looked back at Bailey, obviously avoiding Kyle. "Just go down that hall, the rooms are clearly numbered."

Seeing Kyle was about ready to make a comment, Bailey grabbed his arm. "Let's go find Jess."

The three went down the hall and then stood outside the closed door with the number three on the wall next to the door. Feeling her nerves doing a tap dance, Bailey was afraid to even knock. She kept thinking what if they were all wrong thinking Jess' healing had worked.

Patrick saw the look on her face. "It's okay Bailey, Joshua would have let us know if Jess had problems."

Nodding his agreement, Kyle reached up and knocked on the door. Hearing Jess holler 'come in', a smile lit up Baileys' face. Pushing the door open, she ran over to where Jess was laying on a hospital bed. Other than being a little pale, the man looked like the same Jess she'd said good-bye to when he'd stepped through the doorway headed to the Earth World. As she stepped to the bed, Bailey was torn between wanting to hug Jess in relief and wanting to slap him for scaring her to death. She chose the hug. Sitting next to him on the bed, she bent down and hugged the man, feeling his strong arms wrap around her. Pulling away, she looked into his brown eyes. "I'm so glad you're okay. You had all of us worried sick."
After saying the words, Bailey made good on her second option and slapped him on the

chest. "And if you ever do anything that foolish again, I'll beat the hell out of you."

Stepping up to the bed, Kyle smiled. "And I'll be first in line to help her." The smile turned into a laugh. "Damn good to have you back Jess. How'd it feel being an old geezer?"

Jess shook his head. "The next time someone complains about growing old, you can bet they'll get my sympathy."

Patrick stepped up next to Kyle. "Thank goodness for that healing chamber. You look almost like your old self. How are you feeling?"

Smiling, Jess flexed the hands he'd been staring at just moments before his team had arrived. He'd been thinking of how those hands had looked just before the crash. Then how he had used those hands that, at the time, belonged to a century year old man, to try and claw his way toward the place where he knew the doorway could be opened. From his vantage point on the ground he had seen the slight wavering of the air and knew the door was so close, yet it might as well have been miles away. When he'd blacked out, before reaching his destination, the last thing he'd been thinking was he'd really screwed up and he was going to be taking his last breath alone and in a world where he didn't belong.

Looking back at Patrick, Jess nodded his head. "I'm feeling pretty damn good, thanks to the three of you. Joshua told me what you did. Thanks for saving my ass. I owe you all, big time."

Patrick shook his head. "Bailey's the one who saved you. Kyle and I weren't planning on checking on you for another half an hour."

Taking a chair, Kyle nodded. "Patrick's right, Bailey wouldn't wait and it was a damn good thing too. You didn't look so hot when we pulled you through that doorway."

Taking Baileys' hand, Jess lifted it to his mouth and kissed it. "No wonder I love you so much."

Bailey grinned, but then bit her lip that was trembling. "I love you too Jess."

Wiping away a tear, Bailey shook her head. "Never mind all that. You're good now and that's what counts." Bailey frowned.

"I guess if Joshua talked to you, he told you about the updates on the asteroid."

The brown eyes narrowed as Jess nodded. "He told me. Damn, that whole thing is just crazy. Did all of you get a chance to look at our first World?"

Nodding, Kyle then frowned. "Are they going to let you go with us?"

Jess laughed. "They'd have a hell of a time stopping me. Joshua said I should be released from here in the morning. More because he needs all hands on deck for the traveling than because he thinks I deserve to go. After we find a place for the evacuation of everyone in Terrene I'm sure Joshua Porter will have a few things to say about my trip to the Earth World. The important thing now is for everyone to work together to find a safe haven."

Nodding, Kyle rubbed his stomach. "Before any of that, I need to find some dinner. I think I'm heading to the cafeteria."

Patrick nodded. "Me too, if you're out of here tomorrow, we can fill you in on our new agenda for traveling."
Turning to Bailey, Patrick lifted an eyebrow in curiosity. "You coming with us Bailey?"

Shifting slightly on the bed, Bailey shook her head. "You two go ahead, I think I'll sit with Jess awhile."

Expecting that answer, both Patrick and Kyle nodded. Then, with good-byes to both Jess and Bailey they headed out of the room in search of food.

As soon as they left, Bailey slapped Jess on the hip. "Slide over Jess, I think there's room for both of us on that bed."

Smiling, Jess slid over. As soon as Bailey stretched out next to him, Jess ran a hand through her shoulder length brown hair.
"I really am sorry I worried you. I thought I could make it work, I really did. There's just

something about that world. I don't know what happened to Earth, but I bet years ago it was a paradise."

A shiver ran through Bailey. "It's a long way from that now. I just hope we can find a real paradise and in time to save Terrene." Bailey snuggled closer to Jess. "All I care about right now is that you're okay. Could you do me a favor Jess?"

Frowning, Jess nodded. "Anything, all you have to do is ask."

Bailey sighed. "Just hold me Jess, that's all I want."

Wrapping his arms around Bailey, Jess smiled. "That's one request I'll be happy to comply with."
Feeling herself relax in the warmth and strength of Jess' arms, Bailey closed her eyes. She was tired from the stress of the day. Feeling the safety of her life partners' hug, Bailey let her tiredness overtake her and slept.
Jess, still recovering, did the same.

Chapter 4

The next morning, the feeling of something tickling Baileys' ear woke her up. Without opening her eyes, she slapped at the area, surprised when her hand made contact with something solid. The blue eyes opened quickly then and Bailey rolled over, almost falling out of the narrow bed. Seeing Jess beside her, Bailey shook her head.
"I'm sorry Jess. I forgot where I was."

Jess put a pretend frown on his face. "And here I am supposed to be recuperating too. Hope I don't end up with a black eye. What will the nurses say?"

The woman stepping in the room let out a half grunt. "I think any of the nurses taking care of you and putting up with you would unanimously say you deserved it."

The nurse smiled at Bailey. "Hi, I'm Megan, you must be Bailey."

Bailey smiled. "I am, it's nice to meet you. I hope Jess didn't give you too much trouble."

Shaking her head, Megan laughed. "Oh, I've seen worse. I have good news for both of you though. The doctor says he's going to release Jess. He's making his rounds this morning. He should be headed this way soon."

Jess looked down at the hospital gown he was wearing. "You people didn't happen to save my clothes did you?"

Rolling her eyes, Megan shook her head. "Other than that jacket you were wearing, everything else got thrown out and good riddance. That stuff was ruined."

Bailey stood. "I'll go get you some clothes Jess."

As Jess nodded, he glanced quickly at Megan before turning back to Bailey. "I could just wear the jacket. It's only a short walk to the apartment."

Both Bailey and Megan shook their heads and Bailey laughed harshly. "I think everyone around here has seen enough of your body. I'll be right back."

Bailey smiled as she walked through the Postern facility. It was nice to see the old Jess back, lame jokes and all.

Two hours later, Bailey and Jess stepped into the lab where Kyle and Patrick were seated in front of Patricks' computer system. The men turned at the sound of the door opening. Kyle was the first to speak. "Jess, wow, I can't believe they let you out."

Bailey laughed. "Actually, they were glad to get rid of him."

A smile lit up Patricks' face. "I'm happy to see you Jess. Are you feeling up to all of this?"

Rubbing his hands together, Jess nodded. "I'm more than ready, you want to fill me in on what's happening?"

As the two took chairs, Patrick told Jess what they had found out before turning to the computer and hitting a button. "The best thing is we just got the audio feed. I think it will make Bailey happy."

Looking at Patrick, Bailey frowned. Then as she heard people talking a smile slipped across her face. "They speak English, what a relief. Even if we had years instead of months, I still wouldn't do well learning a foreign language. That is great news Patrick."

Turning up the volume slightly, Patrick looked at the others. "You can hear a minor accent, but it's not too heavy. If we have to talk to anyone, I think we'll be fine."

Both Kyle and Jess were nodding. Jess spoke, adding a slight southern drawl to his voice. "When do y'all think we can head on in to Province?"

Bailey laughed. "That was great Jess, you can do any talking we have to do."
Bailey turned to Patrick. "When will we go in? Remember we have to get our clothes adjusted. They don't dress a lot different than us. I just hope all the women don't

wear those skirts. I don't even think I own a skirt."

Patrick turned to the computer and pulled up a still picture of a woman wearing pants that ended mid-calf. "How about something like this?"

Bailey leaned over to take a closer look. "Much better, the other pictures I saw all had women wearing flared out skirts. This I can do."

Patrick smiled. "Us men can get away with jeans and t-shirts. You're in luck Jess, it looks like the guys in Province like leather jackets. You'll definitely fit in."

Jess nodded. "Yeah, but mine looks a little worse for wear. I can leave it behind, but when are we going in?"

Shrugging Patrick sighed. "I think the first thing in the morning. Are you really sure you're up to this Jess? You've been through a hell of a lot. You looked pretty bad when Kyle pulled you from Earth."

Jess nodded. "I'm okay, I'd tell you if I wasn't. Besides Joshua wouldn't have let those doctors release me if I wasn't in good enough shape."

Patrick nodded. "That's good enough for me. I think we should take it easy the remainder of the day and get some rest. That will let us get together what we need and then check in with Joshua."

Bailey stood. "I'm headed back to the apartment to figure out an outfit. Let's just meet back here in a couple of hours. Then we can head over and talk to Joshua. After that we can grab some dinner."

Bailey left the men and headed to the apartment she shared with Jess. The first two years she'd been with the Postern Project, she'd had her own place. When the two had become a couple and decided to move in together, Joshua had found them the slightly larger place they now shared.
Everyone who worked at the Postern facility also lived here. Not only was it convenient, but Bailey was sure it was also the best way to keep the experimental traveling a secret.

It didn't take her long to find a suitable outfit for travel. She thought about trying to dig something out for Jess to wear, but since all the men could get away with having on jeans and a t-shirt, Bailey decided to let Jess pick out his own.

Instead, Bailey went into the kitchen, made herself a cup of coffee, then sat back at the table and took some time to just unwind. Instead of relaxing, her thoughts returned to Jess and how close she'd come to losing him. Back to when Marna, the fifth member of the Alpha group, had died, Bailey always feared losing another member of their team. Marna had not only been her co-worker on the Alpha team, she'd been Baileys' best friend. It had been so nice to have another female in the group to talk things over with. Bailey knew some of the women on the other teams, but none well enough to call her friend. After Marna died, Bailey didn't want to get close to any other women. She had Jess, Kyle and Patrick as team members and friends. She'd decided that was a big enough circle.

Still, she couldn't help but wish Marna was

still here to share her feelings with. Unlike Jess, who had almost died in a parallel world, Marna had gotten sick right here in Terrene and also had died here. A rare disease invading her body. A viral infection that had roared like a wild fire through the once vivacious young woman. A few short weeks after the diagnosis, Marna was gone.

Bailey sighed and then stood to dump the coffee she'd only half finished. Knowing sitting here wasn't giving her the relaxation she'd had in mind, Bailey headed back to the lab.

When she stepped in, Jess looked at her and frowned, his brown eyes concerned. "Are you okay?"

Sighing, Bailey nodded. "Just caught up in some old memories. Can we just head over and talk to Joshua? I think that will keep my mind occupied."

Stepping over to Bailey, Jess thought he knew exactly what memories had Bailey looking so upset. "We can hold off talking to Joshua. How about I buy you a cup of coffee and we can discuss what's got you feeling down."

Bailey shook her head. "I'm okay, really. Let's go see Joshua and let him know we're ready to travel."

The four members of the Alpha team left the lab and headed to Joshuas' office. Unlike some of the higher up team members, Joshua didn't have a secretary. He liked to work alone and was sitting at his desk when the team stepped in. He looked up almost having to stand to see above the clutter of papers and boxes strewn across the top of the desk. "C'mon in and sit down. I was just getting ready to call you. I've already talked to the other teams. I thought I would leave yours for last. I figured you might need a little extra time, you know, given Jess' recent problems."

Standing up as soon as the others sat down, Joshua opened one of the boxes. "I think you're going to like the new development we've come up with."

Handing each of the members a three inch by five inch rectangular package, Joshua shook his head. "I wish we would have had these years ago."

Frowning at the box that was less than an inch thick, Bailey opened hers first. "What is this?"

Now, Joshua smiled. "That is a device that can find and open doorways. It is also capable of checking the atmospheric conditions in the other worlds. One more thing, it is also a communication device. You can't talk to other worlds, but you can talk to each other while you are in those other places."

Looking at his own device, Patrick frowned. "Why haven't we had these all along?"

Shrugging, Joshua sighed. "To tell you the truth, we just didn't have the technology. Each world, although linked is so different. We actually borrowed this technology from the Earth World. Their technology is so much greater than ours. It always amazes me they haven't opened their own doorways for traveling. They most certainly have the technology to do it."

Jess' brown eyes narrowed. "Maybe they weren't forced into it yet like we were. Knowing you'll be hit by an asteroid forces you to do some amazing things."

Kyle was nodding. "That's for damn sure. So, how do these things work?"

Pulling his own device from the box, Joshua held it up, pointing at the side of the slim article. "First, you push the button you see here on the side. That will turn on your EDAR Unit."

Joshua smiled at the curious looks on the crew members' faces. "The EDAR is an Electron Diffusion and Atmospheric Reader. When magnetic lines cross, the electrons diffuse to create doorways. The EDAR can find those spots. We already pre-check the atmosphere before anyone steps through the doorways, but it's good to have confirmation while you're in the other worlds."

Pointing at the others, Joshua smiled. "Okay everyone, turn on your EDARs and we can get started."

As soon as the four members turned on their devices, Joshua slowly walked them through the operations. The basics were fairly simple and it wasn't long before the Alpha team had the system figured out.

When they finished, Patrick was frowning. "Why can't we use the EDAR to communicate with our base here in Terrene?"

Looking at Patrick, Joshua shook his head. "I wish we could, but passing through the doorway puts up a barrier we haven't got the technology to break through yet. Even with our cameras, they have to be set up on this side of the doorway and then we've been lucky to even have that video and audio feed."

Looking at the team members, Joshuas' face grew serious. "There is something new, we've decide to add to the expeditions though."

The team stared expectantly at Joshua. He stared back, wondering how they'd take this new development in how their journeys would be working. "I'm going to ask you to split up your team. When you go to Province, I want Jess and Bailey to go to one section and Kyle and Patrick to go to another. We've been able to find out they call their separate areas in that world territories. I think by splitting up we can get more done in the short time we have."

All four members of the team were surprised at the request, especially Patrick, who didn't always join in the traveling. Most times he was the one who stayed behind in the lab to monitor the expeditions. A frown covered his face. "Who will be in the lab to be the observer during the traveling?"

Pointing at his own chest, Joshua grinned, green eyes sparkling. "In the case of the Alpha team, I will be the one in the lab. I have others who will take over for the Beta and Gamma teams. I'm guessing the four of you will be okay with me being the one to sit at the computer."

All of the team members nodded, but Jess began laughing. "You know more about all of this than anyone. I think we'll feel safe in your hands."

Joshua nodded. "I'm glad to hear that. Now, the important question, will you four be ready to travel in the morning?"

Jess answered for all of them. "I feel confident we're all ready. If we split up in the two teams, where do we travel to? The video I saw only showed a diner and the area just around it in Province."

Joshua nodded. "We have another camera set up in front of a library building. We're hoping that will be the best place to gather information on Province, just like you did in the Earth World. Hopefully they have a computer system you can take information off of." Joshua sat down. "If you feel prepared, then I suggest all of you get a good nights' sleep. We can meet in your lab first thing in the morning."

Leaving Joshuas' office, Kyle and Patrick headed to the cafeteria for dinner. Bailey and Jess turned down their friends' invite to join them. The two preferred to head to their apartment for a quiet dinner and to spend time together after Jess' prolonged absence and miraculous return.

Chapter 5

The next morning, as the four members of the Alpha team entered the lab. Joshua was already there and seated at the computer. "Glad you could make it."

Kyle laughed. "What did you do, sleep here?"

Shaking his head, and pushing back the sandy hair, Joshua smiled, the green eyes looking tired. "Not quite, but I have been here an hour or two. I have the video of the second site in Province for all of you to look at before you head through the doorway."

Grabbing chairs, the four moved them over where they could all see the computer screen. The scene in front of them was void of people, although a few cars were parked in front of an older brick building. Joshua

pointed at the screen. "This building is typical of what you'd find elsewhere in Province. Their technology, I'm sorry to say, is far behind our own."
Joshua turned to look at Patrick and Kyle. "I'd like the two of you to travel here. I know Patrick is the technology wizard of the group. I'm hoping you can get the valuable information we need from whatever computers Province uses."

Patrick frowned. "What if they don't have the technology to even have computers?"

Joshua nodded. "I thought about that, and it is a definite possibility. Just get any information you can, any way you can." Turning in his seat, Joshuas' gaze went to Bailey and Jess. "You two are headed to the diner. Talk to the people, try to get a feel for them and the place. We need to know about their government and their laws, or lack of. Also try and find out what kind of work the people do, how they live, things like that. This first trip is a fact finding mission. I've been in contact with the Chamber of Guardians, they know about the new asteroid trajectory and time table. They're anxiously waiting for us to find a new world before they go public with the new information. I don't have to tell you how eager they are to hear what results your team and the others can bring back."

The team members nodded, knowing the governing body from Terrene, the Chamber of Guardians, would need that information to save the World of Terrene and its' people. Jess stared at Joshua. "We'll do our best. I think we should get started."

As the four stood, Joshua noticed their clothes. He hadn't paid any attention earlier to the altered outfits. "I like the clothes by the way. I think the four of you especially Bailey, look great. I like the ponytail."

Bailey smiled, putting a hand back to touch the ponytail that sat high on the back of her head. Smiling back, Joshua continued. "I guess if you are all ready to start, I'll open the doorway for Patrick and Kyle first." Joshua pointed at the two men. "Do you have your EDAR Units?"

Both Kyle and Patrick patted their back pockets where they'd put their units and nodded.

Joshua smiled. "Good, step over to the door and we'll get you on your way."

As they approached the door, Kyle smiled at Patrick. "It's been a while since we traveled together. Glad to have you as my side kick."

Patrick laughed. "Same here."

As the door opened the two men turned around and held up a hand in a good-bye to the others in the room. Then, turning back, stepped through the doorway.

Joshua turned back to the computer in time to see the two men heading for the library building through the viewer they'd set up. Wishing them a silent 'good luck', he hit the commands on the computer to prepare for the other doorway.
Turning to Jess and Bailey he grinned as he saw the two had joined hands.
Looking back at the computer screen, Joshua watched the front of the restaurant.

He waited a moment before opening the doorway, waiting for a couple in Province to get out of their car and go into the restaurant. As soon as they went inside, he turned to face Jess and Bailey. "The coast is clear, get ready to travel."

Joshua hit a button and then turned to watch the second half of the Alpha team step through the doorway.

As the door closed behind them, Joshua whispered. "Smooth travels", before turning back to the computer. He brought up a split screen so he could watch all four members of the team that had been his first to put together years ago when he had started Project Postern.

He wasn't able to watch any of them long. Patrick and Kyle disappeared from his view into the library, leaving Joshua with just a picture of the libraries closed front doors. Turning to the other side of the split screen, Joshua watched Bailey and Jess. The two stepped to the side of the diner instead of going in through the front glass doors. Frowning, Joshua wondered what the two were up to as they slipped out of his sight.

*   *   *

As Bailey and Jess walked to the diner, they heard a scream. Turning to Jess, Bailey frowned. "That sounded like a woman. Oh hell Jess, I think someone's in trouble."

Grabbing Baileys' hand he had let go of while they were walking, Jess nodded. "The sound came from behind the building. C'mon, let's go see what's going on."

The two hurried along the side of the building, while they tried not to be noticed. As soon as they got to the end, Jess stopped and motioned for Bailey to do the same. Letting go of her hand, Jess carefully peered around the back of the building to try and see what was happening.
As he took in the scene, playing out in what looked like a parking lot, he quickly ducked back and whispered to Bailey. "You were right, that was a womans' scream. I don't know what the hell is going on. She's on the ground with her hands behind her back. I think they're tied behind her. There's a

young girl kneeling beside her. A man is standing next to them. He has some kind of club in his hand. From the looks of that woman, he's hit her more than once."

The blue eyes grew wide. As Bailey shook her head, the dark ponytail swung from side to side. "What are we going to do Jess? We can't let that guy beat on a woman, no matter what his reasons are. Is the girl hurt?"

Jess shook his head. "I don't think so, not physically anyway."
Jess looked along the side of the building in search of something he could use for a weapon. When he didn't see anything, Jess sighed. "I'm going to have to go out and try to stop him. While I keep him busy, you head to the woman and the girl. Try to get them out of there."

Bailey shook her head. "Jess, no, that guy has a club and maybe even other weapons. We never carry weapons."

Jess nodded. "Yeah, don't remind me. I'm hoping when I step out there, that guy will be surprised to see me and hopefully that will give me some kind of advantage. I'm sorry Bailey, I just don't see any other way."

Bailey was going to argue, when another scream cut through the air. Baileys' face covered with anguish, but Jess' brown eyes hardened. He glanced at Bailey. "Just be ready."

Stepping out into the area behind the building, Jess yelled. "Hey, what the hell are you doing?"

The man who turned to glare at Jess was at least three inches taller than Jess and a good fifty pounds heavier. The voice came out with what sounded like a southern drawl. "None of your damn business. If you have a lick of sense, you'll turn around right now and get out of here."

Taking another step toward the man, Jess shook his head. "I'm afraid I can't do that. It looks to me like that lady and the little girl are at a disadvantage. I have to tell you, that's something I just don't like to see."

Turning away from the woman and child, the man scowled at Jess. "Doesn't matter to me what you think. I'm in charge of the Grandview Territory. I don't need to explain myself or my actions to you or anyone else for that matter."

Taking a quick glance back at Bailey, Jess could see she was watching the woman and the young girl. He knew she would be waiting for the chance for a clear shot, so she could get to them. Jess moved further away from the building, hoping to get the man to move away from the woman, still kneeling on the ground. Slowly, he stepped in a half circle. The big man, seeing Jess moving away, turned more toward him than the woman.

As she watched the two men, Bailey moved quickly. Out of the corner of his eye, Jess saw the movement and continued talking, trying to divert the mans' attention. Jess held up his hands. "I don't want any trouble. I just can't see a big guy like you beating on a woman, and having a little girl watch."

The mans' dark eyes narrowed. "She's a law breaker, and none of this is any of your damn business. If you don't want to bring on your own punishment, I suggest you just move on."

Bailey had reached the woman and Jess could see her bending down to touch the woman's shoulder.

While Jess continued trying to speak to the man, Bailey spoke to the woman, keeping her voice quiet. "Can you walk? We need to get you away from here. Maybe inside the diner so that I can untie you."

The woman shook her head and looked up at Bailey. The blue eyes were red rimmed from crying and wide with fear. The side of her face had a red mark that was already beginning to turn into a bruise. "Not the diner, they'd just as soon kill me."

Sighing, Bailey grabbed the womans' arm and pulled her to her feet. The lady was even shorter than Baileys' five foot two inches of height and felt to Bailey like she was at least twenty pounds lighter. As soon as the woman stood, so did the child. Bailey looked at the girl, who looked so much like the woman, she knew that had to be mother and daughter. "What's your name honey?"

The girl looked even more frightened than the woman. She glance quickly at Bailey before looking down at her feet. "My name's Miya."

Nodding, Bailey smiled. "Hi Miya, I'm Bailey. I'm here to help. Is this your mom?"

When Miya nodded, Bailey let out a sigh. "I'm going to help your mom out of here. Can you keep up with us?"
Again, Bailey saw the head bob. "Okay then, we need to move fast. Are you ready?"

Glancing quickly at her mom, the girl nodded. Taking the womans' arm, Bailey pulled her to the side of the building, glad to see Miya sticking close. She was also relieved that she didn't hear any objections from the man, who was still having a heated discussion with Jess.
As they reached the building, Bailey turned to look back and check on Jess. He was

facing the man, but walking backwards, one slow step at a time. Each stride took both men further from the diner.

Bailey turned to the woman. "What's your name?"

A trembling voice answered. "Lucy, my name is Lucy Sherwood, thanks for your help."

Bailey shook her head. "Never mind that, we need to get you out of here. Turn around and let me get that rope untied."

Turning, Lucy felt Bailey pulling at the knotted rope. "I'd like to get out of here, but there's no where I can go. That man's the Regulator for Grandview territory. Even if I headed for another territory, I'd be back here for punishment."

As Bailey got the rope untied, Lucy turned to face her. Bailey stared at the woman. "What did you do anyway?"

Bending down to pick up Miya, Lucy shook her head. "I didn't have my food card to pay for our lunch. I thought I had it, but I must have lost it."

Bailey didn't have time to find out what a food card was. Instead, worried about Jess, she pulled her EDAR Unit from her pocket and spoke into it. "Kyle, Patrick, are you there? Can you hear me? Kyle, Patrick, Are you there?"

* * *

In the library, Kyle and Patrick were having their own problems. Neither man was in danger, like Bailey and Jess, but they were disheartened to find the World of Province not only had less technology than Terrene, they had almost no technology at all. The girl at the front desk of the library had acted like the word, computer, was part of a foreign language.

After telling her a made up story about being writers needing information on the history of Province, Kyle and Patrick had been directed to the back of the musty smelling library. They were seated at a table with stacks of history books scattered across its' surface when the sound of Baileys' frantic voice was heard coming out of the EDAR Units they both carried.

Patrick pulled his unit from his pocket first. "Bailey, we're here, what's wrong?"

Blowing out a relieved breath, Bailey ignored Lucys' stare of amazement and spoke into her unit. "We're in trouble. Jess is behind the diner arguing with a guy who's some kind of Regulator. I don't have time to explain. We need your help."

Hearing the words, Kyle stood. "We're on our way Bailey, Just hang in there."

Bailey slid the EDAR back into her pocket and then seeing the curious look Lucy was giving her, tried to figure out first, how to explain, and second, how they were going to get out of this mess.

*   *   *

Rushing out of the library, Kyle turned to Patrick. "How do we get to that diner? I don't even know how far it is from here."

Patrick shrugged. "There's no way to tell. I think our only option is to take the doorway back to the lab and then re-enter Province from there and head to the diner."

Kyle nodded. "Let's do it. From the sound of Baileys' voice, things aren't good."

As they crossed the road, Kyle held up his EDAR Unit and pointed it toward where he could see a slight distortion in the air, twenty feet ahead. As he pushed a button, the doorway materialized and opened. Kyle and Patrick stepped through, the door closing behind them.

*   *   *

In the lab, Joshua had watched the two men emerge from the library and rush toward the doorway. On the other screen, Joshua saw no sign of the other half of the Alpha team. A fact that had his nerves on edge and his stomach in knots. Now though, his attention was riveted on the doorway in the lab.

As it opened and Kyle and Patrick stepped in, Joshua stood and walked over quickly to meet them. "What's going on? Why are you two back already?"

Kyle shrugged. "We need to get to that diner. Jess and Bailey are in trouble."

Patrick saw the confused look on Joshuas' face. "We got a message from Bailey. We don't know what's happening, but we need to get out there."

Kyle nodded. "And we need some type of weapons."

Ignoring the frown that had appeared on Joshuas' face, Kyle ran to a cupboard and grabbed out two black tubes. Stepping back over, he handed one to Patrick, then turned to Joshua. "We don't know what's happening. These stunners won't kill anyone, but if we need to, they'll put them out of commission for a few minutes at least."

Joshua was shaking his head. "I don't like any use of weapons Kyle, you know that."

Kyle nodded. "I do know that, but our teammates are in trouble. Are you going to program that doorway so we can go help them are not?"

Sighing, Joshua nodded. "Just be careful." Going to the computer, Joshua pulled up the program that showed the diner. As he stared at almost the same scene he'd been watching since Jess and Bailey had entered Province earlier, he activated the door way.
Turning, he watched Kyle and Patrick step through. As the door shut behind them, Joshua looked back anxiously at the computer screen.

*   *   *

As the two men stepped into Province, Kyle pointed at the diner. "Bailey said Jess was behind there. Let's go up the side of the building, maybe we can sneak to the back without being seen."

Taking a tighter grip on his stunner, Patrick nodded. The two men headed to the side of the restaurant and then started walking cautiously. They hadn't gotten far when they saw Bailey standing on the side of the building with a woman holding a child.

At virtually the same time, Bailey saw the two men and motioned for them to join her. Kyle reached her first. Glancing at the woman and child, he frowned before turning to face Bailey. "Where's Jess?"

Bailey pointed toward the back of the building. "He's back there, but he's not alone. Some guy, who calls himself a Regulator, was beating on this woman. I got her and her daughter over to here, but Jess is still back there with that thing."

Kyle nodded. "I want you to stay here with these two."
Turning to Patrick, Kyle grimaced. "You ready for this?"

Patrick nodded. "Let's go get Jess."

Leaving Bailey with the woman and child, the two men jogged to the end of the building. Looking around the corner of the building, Kyle saw Jess and another man. The two were on the ground, with the stranger on top of Jess. His arm was in the air, his hand clenching a club.

Fire in his eyes, Kyle let out a scream and ran toward the men. Instead of pulling out his stunner, Kyle rushed over and putting down his head, barreled into the larger man. The motion knocking him off of Jess.

As Kyle and the other man hit the ground, Patrick ran over and pushed his stunner into the man's side. Even after the guys' body started jerking, Patrick kept up the pressure until the mans' eyes rolled back in his head and he fell over to lay flat on the ground.

By then, Jess had gotten up. He stared at the man, then at Patrick. "You didn't kill him did you?"

As he reached down to touch the mans' neck and felt a weak pulse, Patrick shook his head. "He's alive, but he may be out for a while."

Kyle too had made it to his feet. He placed a hand on Patricks' shoulder. "Thanks for backing me up. When I saw that guy on Jess, I lost it. I forgot all about my stunner."

Giving Kyle a grin, Patrick shook his head and then his face turned serious. "I think we should all get out of here before that guy wakes up. Let's head back to the lab. I think we've learned all we need to know about Province."

Kyle nodded. "And none of it good."

Rubbing a cut on his face, Jess looked at his two friends. "Thanks for saving my ass, yet again."

Kyle laughed. "I hope you don't plan on that happening with every trip you make. It's really getting to be a bad habit Jess."

Nodding, Jess laughed. "You don't have to remind me."

Jess then frowned. "Where's Bailey anyway? The last time I saw here, she was hauling that lady and the little girl out of here."

Kyle nodded. "All three are waiting on the side of the building."

Jess' face filled with confusion. "What are the other two still doing here? Bailey should have sent them on their way by now."

Kyle shrugged. "Maybe you should ask her Jess."

Nodding, Jess started walking toward the building. Turning back, he looked at Kyle and Patrick. "You two coming?"

With smiles on their faces, and one last look at the man on the ground, the two joined Jess. As they walked toward the side of the building, Bailey poked her head out.

She'd been standing around the corner, torn between wanting to look, and afraid to, because she knew if Jess was in trouble, she'd have a hard time keeping herself from running out to help. Her responsibility, like it or not, right now, was figuring out what to do with Lucy and Miya. Still, she couldn't stand not knowing and had finally taken a look. Seeing the three men, Bailey forgot her responsibility and ran out to hug Jess. "Are you okay Jess? I'm so sorry, I didn't know what to do."

Holding her tight, Jess smiled. "I'm fine and it looks like you knew exactly what to do."

Kyle nodded. "Yeah, call us, the all-important back up men."

Jess rolled his eyes. "I hope next time it's me that has to save you."

Laughing, Kyle shook his head. "I don't see that happening. But hey, keep dreamin'."

Ignoring Kyle, Jess tuned to Bailey. "Where's the woman and the little girl? Are they okay?"

Bailey nodded. "They're over on the side of the building. We may have a problem though. Lucy told me, oh, that's her name, by the way. The little girl is her daughter, her name is Miya. Anyway, Lucy said she has nowhere safe to go. Even if she could leave this territory, that Regulator you three left on the ground, would find her."

Jess frowned, not liking what he had a suspicion that Bailey was thinking. He decided not to say anything until he talked to Lucy first. "Okay, don't worry Bailey, we'll figure this out."

The Alpha team went around the building where a scared looking Lucy held a sleepy Miya. Bailey went ahead of the men over to Lucy. "It's okay, these are my friends." Bailey turned and pointed at each man as she introduced them. The men only nodded and smiled, unsure what to say to the woman.

Finally, Jess took a step closer. "You don't need to be afraid, we aren't letting that big goon hurt you anymore. By the way, how are your injuries?"

Lucy shrugged her shoulders, an awkward gesture holding a five year old. "I'm okay, although by tomorrow, I think these bruises are going to hurt like hell. I just want to say thanks to all of you for helping me."

Behind Jess, Kyle frowned. "What was all that about anyway? What's wrong with that guy?"

Lucy laughed harshly. "What he did is normal punishment for stealing."
Then she shook her head. "I wasn't actually stealing, Miya and I stopped here to eat. I didn't realize until we finished that I didn't have my food card with me. The people working at the diner decided I was trying to get a free meal and called in Regulator Thompson. He forced me to go out back where he tied my hands and began beating me with that damn club of his."
Lucy smoothed Miyas' hair as she talked. "I just wish Miya didn't have to witness all that."

Miya laid her head on Lucys' shoulder and closed her eyes. Jess' brown eyes narrowed. "Bailey says you don't have a safe place to go. Don't you have family or friends that could help?"

As Lucy shook her head, the others were surprised to see tears in her eyes. "My husband is dead. If I have family anywhere, I don't know about it. Since Adam passed away, it's just been Miya and me."
Lucy didn't add that Miyas' father had died two months before the little girl was born.

Bailey looked at Jess with sympathy in her blue eyes. "We have to take them with us Jess. We can't leave them here in this awful world."

The stare Bailey received was filled with disbelief. "Bailey, do you realize what you're saying? You know why we're here. Bringing people back is the exact opposite of our goal. Even if we did, what happens in six months if we can't achieve that goal?"

The blue eyes hardened. "Even if we can only give them six months, that's better than what they'd get from that Regulator here. I'd say six days with him would be a stretch."

Jess was about to object again, when he was left speechless by Patricks' words.

"Bailey's right Jess. They need to come with us. I don't think I could live with myself if we left them here to fend for themselves."

Throwing his hands in the air, Jess turned to Kyle for help. He knew he wouldn't be getting any when Kyle laughed. "Sorry Jess, I'm with them, but only on one condition." He turned to Bailey. "You explain all this to Joshua."

Bailey hugged Patrick first and then stepped over to pull Kyle in a hug. "I'll talk to Joshua. Now, can we just get the hell out of here?"

Bailey turned to Lucy, who was staring wide-eyed, wondering what in the world was happening. Placing a hand on the womans' shoulder, Bailey tried to give her a reassuring smile. "I knows this is probably all really confusing, but I promise, we'll help both you and Miya. If you don't like the place we're headed, we can always bring you back."

Shaking her head, Lucy stared at Bailey. "There's nothing here for me and I don't want Miya growing up in this place either. I trust you. Although the four of you have already done more than I expected. I just don't know how to thank you."

Bailey shook her head. "We don't need thanks and you better wait for all of that until we take you to our home. You might not like it."

Patrick stepped over. "Why don't you let me carry your little one? After what you've been through, you're probably worn out."

Grateful for the offer, Lucy carefully handed Patrick the sleeping child and then the small group headed for the doorway.

Chapter 6

Back in the lab, Joshua watched in disbelief as his Alpha team walked toward the viewer, on their way to the doorway, accompanied by not only an extra woman, but he was sure Patrick had a child in his arms.

Standing, Joshua walked to the doorway, before it opened. Then he stood there, arms folded, waiting for the team and an explanation.

As the doorway opened, Bailey was the first to step through. Almost running into Joshua, Bailey drew back slightly, blue eyes wide. "Joshua, you scared me. I thought you'd be over at the computer."

The eyebrows lifted above the green eyes. "I was at the computer, in fact, that's why I'm standing here now. What in the hell is going on?"

The others stood behind Bailey, only moving far enough to allow the doorway to close behind them. Jess then stepped up next to Bailey, but looked at Joshua. "Maybe we could all take a seat, before we begin explaining. We've been through a strange ordeal."

Jess turned to stare at Bailey a moment. She was sure that she could see a hint of laughter in his dark eyes. "Bailey has some things she'd like to tell you."

Frowning, Joshua pointed at Jess' face. "You have quite the scratch on your face Jess. Patrick and Kyle said you were in trouble, is that part of it?"

Looking over at the woman who had come in with the Alpha crew and then over at Patrick, carrying a child. Joshua sighed, not waiting for Jess to answer him. Instead he pointed at one of the lab tables that had some chairs sitting close by.

"Okay, let's sit down. I really want to hear this explanation."

As Joshua turned away and headed for the table, Bailey turned to Jess and whispered. "Thanks a lot Jess."

Nodding, Jess smiled. "Anytime Bailey, always glad to help."
He turned to Lucy. "C'mon over and sit down. When we're finished here, we'll take you down and have those cuts looked at."

Lucy frowned, but joined the others as they pulled up chairs and joined Joshua at the table. As soon as they all had taken their seats, Joshua stared at Bailey. "What's going on Bailey?"

Looking at the others first and finding no help in their faces, Bailey turned back to Joshua. "It's kind of a long story. You probably don't have the time."

Sitting back in his chair, Joshua again folded his arms across his chest. "I'll make the time. I think this is one story I definitely want to hear. Now, what happened out there?"

Knowing she had no other option, Bailey sighed and began talking. The words came out slowly at first, but then speeding up as she told the story. As she began telling Joshua her reasoning behind bringing Lucy and her daughter Miya back with them, Baileys' words were almost running into each other.

To Joshua, it felt like Bailey had to get the story out before he had the chance to raise any objections. As he listened to her stumbling retelling of the story, Joshua had to fight to keep the smile off his face. The one that wanted so desperately to make an appearance. Instead, he turned from Bailey to look at the woman that all this was about. Lucy looked to be in either her late twenties or early thirties. She was pretty in a pixie-ish sort of way. Her most stunning feature had to be the large blue eyes, which stood out in sharp contrast to her dark brown hair. At the moment, those eyes were staring at Joshua with great curiosity. He also noticed now, the bruising on her face, much worse than the damage on Jess'. Knowing that since Lucy and Miya had already been brought through the doorway and into Terrene, his decisions were limited, Joshua turned back to Bailey. "For now, I think you need to take Lucy and Miya down to see the nurses. Get them both checked out. They've been through a traumatic experience. While you three are down in the infirmary, I'll see what I can do about finding them an apartment to use. It looks to me like both of them are going to be needing some clothes also."

Bailey smiled. "Does that mean they'll be staying?"

Joshua nodded. "I don't think we have any other choice. I can't send them back to Province knowing what awaits them in that place. Your bringing them through that doorway exposed them to our project. I don't want them to feel like prisoners, but for the time being, I'd feel better if they weren't given the chance to tell this story to anyone."

Joshua sat forward in his chair, bringing him closer to Bailey. "You must realize that you still have a lot of explaining to do. I understand why you made the decision to bring Lucy and Miya here. At least as much as anyone can comprehend. The two people who are the most entitled to an explanation and some answers are Lucy and Miya. Not just why they are here, but I'm afraid you are going to have to fill them in on the Postern Project, what we are doing and why we only have six months to do it."

The blue eyes grew wide. Bailey hadn't really given any thought to Lucy finding out about this world. She'd only been concerned about removing the woman and her child away from their own world and then to safety. Looking over at Lucy, Bailey wondered if the woman realized the doorway they had all just come through had taken them all into this, a parallel world. Turning back to Joshua, Bailey exhaled a deep breath and then nodded slowly. "I guess I really do have some explaining to do at that. Would it be okay if I took Lucy and Miya down to the infirmary now?"

Joshua nodded, but then frowned as he saw Patrick standing up, still holding Miya in his arms. Looking up at Patrick, Joshua saw a tenderness in the mans' eyes as he looked down at the still sleeping girl. "Why don't you let Lucy take Miya? Bailey can escort them down to the infirmary. She's going to be having a long talk with Lucy anyway. I want the rest of the Alpha team to stay here. Province didn't work out, but I have a new plan I want everyone to look at."

Joshua turned back to Bailey. "I want you to stay with Lucy and Miya. When I find them that apartment, I'll come looking for you."

Bailey turned to Lucy. "Let's go get you looked at."

Nodding, Lucy moved closer to Patrick, who was gently touching Miya. "Wake up honey, your mom's ready to head out."

As Miyas' eyes opened she smiled at Patrick, before reaching out her arms towards Lucy. Taking her daughter from Patrick, Lucy smiled. "Thanks for giving me a break."

A smile put a glimmer in the brown eyes. "You don't have to thank me, I was glad to do it."

Turning from Patrick, Lucy stared at Joshua. "Thanks for letting us stay. I don't have any money to repay you, but I'm a hard worker and if you could find me a job around here, I'd be glad to try and get even."

Joshua stood. "We're here to help. Our mission isn't the one Bailey took on today, but the basics are the same. Welcome to the Postern Facility Lucy." He turned to Miya. "Welcome to you also Miya." As the young girl let out a giggle, Joshua returned his gaze to Lucy. "I'll be in to see you and Miya later, maybe then we can sort some things out."

Nodding, Lucy swiveled away from the men and followed Bailey from the room, unsure what was actually happening, but thinking it sounded more interesting and a lot safer than life in Grandview territory on Province.

*   *   *

As soon as the two woman left with Miya, Joshua stared at the rest of the Alpha team.

"It seems your team is destined to cause me headaches. First Jess, and his unscheduled trip, and now Bailey, rounding up strays. I hope we can focus back on the project now."

Jess sighed. "I'm sorry Joshua. None of us could think of any other way to keep Lucy and Miya safe."

Shaking his head, Joshua sighed. "Never mind, we'll work it out. From the story Bailey told, Province isn't an option for those two anyway."

Now Patricks' head was shaking from side to side. "It was worse than Bailey told you. At the library Kyle and I went to, they didn't even know what a computer was. The history books we were looking through didn't paint a very good picture of that worlds' past either. A few men using fear and punishment to keep the people under their martial law."

Shaking his head, Joshua blew out a breath. "We scratch Province from the list then. We need the kind of leaders who are willing to let our population join theirs and before that damn asteroid hits."

Pointing over toward the computer, Joshua grinned. "Never mind, I have a new place lined up. Don't forget, the other teams are busy searching too. I haven't had word on their first journeys yet, but since you guys are back early, I wasn't expecting any. We might as well get busy."

The men followed Joshua over to the computer and took seats. Joshua sat at the chair in front of the workstation and began typing. A few minutes later, he pulled up a video feed. "We were able to get a camera set up to view into this world. We haven't done much more than that though."

Joshua pointed at the screen where a picture of what looked like a large mansion could be seen. "The world is called Scintilla. From what we can tell, that building is actually a hospital. The researcher who set up the camera went through the doorway and talked to a man out on that grassy field you see. The man spoke our language and said he had a sister in the facility. He called it the Serenity Center. That's about all we have. Other than the fact that the few people we've been able to observe dress about the same as we do. As with all the worlds you travel in, the atmosphere is compatible to our own."

Jess pointed at the screen. "At least it looks warmer than the Earth world. By the looks of those trees, I'd say spring has hit in Scintilla."

Kyle nodded. "Look at those blossoms, I can almost smell them from here."

Patrick frowned. "It almost looks too beautiful. When I see that, I always wonder what horrors are hiding behind the beauty."

Joshua laughed. "And I thought you were the optimistic one."

* * *

Standing next to the hospital bed, Bailey watched the nurse bandage Lucys' cuts and bruises. Miya sat cross legged at the foot of her bed, blue eyes wide as they watched her mom.

The nurse finished and smiled at Lucy. "You're probably going to be sore for a day or two. Would you like something for the pain?"

Lucy shook her head. "No thanks, I'm feeling okay. Thanks for fixing me up."

Shaking her head the nurse smiled. "That's my job. I have a couple more patients to look in on, but if you need anything, just push that button hanging on your bed."

After the nurse left, Bailey sat on the chair closest to the bed. She pointed at the bandages. "I'm sorry we didn't find you soon enough to prevent that."

Raising her hand, Lucy touched the largest bandage on her cheek. "It would have been a lot worse if you hadn't shown up."
The blue eyes narrowed as Lucy brushed back her curly brown hair. "What is this place? I heard it called the Postern Facility and something about a project, but what do you do here?"
Lucy laughed. "Maybe I should also ask

where here is first, I take it we're not in Province anymore."

Taking a deep breath, Bailey blew it out slowly. "The Postern Facility was built for the Postern Project. Joshua Porter is in charge. In fact, what is being done here, Joshua and a man he knew, invented. Using Joshuas' invention, three teams are able to travel to Parallel Worlds. The world you are in now is called Terrene."

Lucy's forehead creased into a frown. "Wait a minute, what are Parallel Worlds?" Seeing Miya staring at her, Lucy held out her arms. "C'mere Miya, I could really use one of your big old hugs."

Miya giggled as she crawled across the bed and on to her mothers' lap. Lucy smoothed down the mass of almost black curls, so like her own, wondering what kind of mess her hair must be in. She turned to Bailey and waited for an answer.

Wishing the others were here to help her explain, Bailey half shrugged. "Joshua could explain a whole lot better than me."

Saying that made Bailey want to slap the man for making her do this alone, but she had to admit all of this was her fault to begin with. "There are countless Parallel Worlds. Most have some similarities, but some are so strange and foreign to us, that going there would be almost impossible. Joshua has researchers whose only job is to look for the worlds that are closest to ours in atmosphere. Unless we can breathe the air without some kind of special equipment we haven't tried to go to those places."

Lucy frowned. "Why are you traveling to the worlds? Don't you like this one?"

Bailey sighed. "I love this world. It's my home. When the traveling began, it was to make contact. We wanted to learn about the other worlds and let them learn about ours."

Again a frown creased Lucy's forehead. "You make it sound like you don't do that now. You said it was to make contact, does that mean you have new reasons now?"

Surprised at Lucys' intuitiveness, Bailey nodded. "Joshua learned that an asteroid was

114

predicted to come close to Terrene. Recently we found out we are in its' direct path."

Lucy stared at Bailey. "What does that mean?"

Silent a moment, Bailey hated to even say the words out loud. Feeling that somehow sharing the facts would seal Terrenes' fate. She bit her lip as she struggled with her thoughts. Finally Baileys' blue eyes locked on the blue eyes staring back at her. "This planet will explode upon impact. No one could survive."

Lucy gasped, holding Miya tighter. "When?"

Shaking her head, Bailey looked away for a moment. As she turned back to Lucy, Baileys' eyes reddened. "We have six months."

Silence filled the room as the two women thought about Baileys' answer. Finally Bailey sighed. "I'm sorry I brought you here only to have to tell you that."

Lucy frowned. "But you're traveling right? Won't you find a place to go? That's the mission Joshua talked about right?"

Bailey nodded. "That's why we were in Province. It didn't work out too well." Seeing the look on Lucys' face, Bailey reached over and patted her leg. "We'll find a place. Right now we have the three teams looking and Joshua is lining up the worlds for us to travel to. At least we have the opportunity of being able to travel. I'm sure other worlds have faced the problem we have and with no options."

Lucys' lips set in a determined line. "I think you'll do it. Don't feel guilty for bringing Miya and me here. We would have had it a lot worse back in Province. You did the right thing."

Letting out a breath, Bailey smiled. "I don't know if I did the right thing, but thanks for saying I did."

A knock on the door, had both women turning to see who was there. Joshua stepped in. He smiled at Lucy. "Looks like the nurses have you fixed up. Would you and Miya like to see your apartment now?"

Smiling, Lucy nodded and slid out of bed. She took Miyas' hand. "We're going to have a new place to live. Do you think you'd like to go look at it?"

Miya nodded, but then frowned, her blue eyes looking worried. "I don't have my doll."

Joshua knelt down so he was eye level with the small girl. "Why don't we go look at your new place and then Bailey can take you to get a new doll."

Miya smiled broadly. "Really, can she have yellow hair?"

Joshua laughed. "I think that can be arranged." He looked over at Lucy. "We better get the two of you some clothes too."

Lucy shook her head. "You can't do that. We can just wash what we're wearing."

Shaking his head, Joshua pointed over at Bailey. "When Bailey brought you here, she didn't have the chance to bring your belongings. The least we can do is make sure you have some clothes and a few toys for Miya."

As Lucy started to shake her head, Joshua held up his hand. "I have to warn you, I'm not the type to take no for an answer."

Bailey laughed. "You better just go along with his plans Lucy. Joshua is known to be pretty stubborn."

As they stepped out into the waiting room, they saw the rest of the Alpha Team sitting in the chairs. Patrick was the first to stand. He went over to Miya and picked her up. "Are you doing okay?"

Miya nodded. "I'm getting a new doll and some clothes. Mom and I are getting a new home too."

Patrick looked over at Joshua, who only shrugged. Returning his attention back to Miya, Patrick smiled. "Good for you." Putting Miya down, Patrick turned back to

Joshua. "Where are you going to get clothes and toys?"

Joshua laughed. "I'm not, this is Baileys' project. After we go take a gander at Lucy and Miyas' apartment, Bailey will be driving these two into town for some shopping."

Hearing Joshuas' statement, Jess stepped over and gave Bailey a hug. "Not too bad of a punishment, going shopping."

Rolling her eyes, Bailey poked Jess in the ribs. "Very funny, just remember, I'm not the only one who's in trouble around here." She turned to Joshua. "So, where's the apartment?"

As Joshua led the way, the others followed. Grabbing Patricks' hand, Miya skipped down the hallway. Lucy smiled at her daughter. "Looks like you found a new friend." Then she turned to Patrick. "She's usually shy with strangers. If she's bothering you, just let me know."

The blond ponytail swung back and forth as Patrick shook his head. "I don't mind at all. Really, I think she's a cute kid."

Lucy laughed. "Most of the time. Thanks for saying so."

As he felt himself wanting to grab Lucys' hand, Patrick only nodded.

Behind the three, Bailey walked with Kyle and Jess, a smile on her face as she watched the interchange.

After the group walked through the two bedroom apartment and listened to Miyas' delighted squeals when she saw someone had decorated her bedroom in bright pinks and purples, the men left to return to the lab. While Bailey, Lucy and Miya, using Joshuas' car, drove away from the Postern facility and headed to town for their impromptu shopping trip.
After two hours of shopping they filled not only the cars' trunk, but most of the backseat with bags of clothes and toys.

As they headed back to the lab, Lucy shook her head. "I'm going to owe Joshua a fortune. I can't believe all the stuff we ended up with."

Not one to usually enjoy shopping, Bailey had to laugh almost in wonder. "It was actually more fun than I thought it would be. We don't get the chance to leave the facility very often. So this trip was actually a treat for me too. The facility is almost completely self-sufficient. We have our housing, medical care, a cafeteria, a work out center and of course, our lab. When we do need something we can't find here, we usually just order it on line."

Turning to look at Bailey, Lucy frowned. "Wait a minute, what's on line?"
The only line Lucy could think of was the one she had set up behind her place in Province for hanging out her laundry.

Bailey had to raise an eyebrow at Lucys' question. She'd forgotten how backward Province was in their technology. "It's shopping on the computer. When we have time, I'll teach you about it."

Realizing she had a lot to learn in this new world, Lucy only sighed.

As Bailey drove, an idea came to her. "You know the person who knows more about computers than anyone is Patrick. Maybe he could teach you better than I could."

Picturing the handsome man with his long blonde hair and compassionate brown eyes, Lucy grinned. Then she shook her head. "I wouldn't want to bother him."

Now it was Bailey who grinned slyly. "Oh, I don't think he'll complain. There's a lot Miya can learn too. She's probably about the right age to be starting school. I have a feeling Patrick will be more than happy to teach both of you."

Chapter 7

Joshua waited three days to give Lucy and Miya a chance to get settled in before he asked the Alpha Team to meet with him in their lab.

Miya and Lucy stayed in their apartment, where they'd been given a computer to practice with while their teacher, Patrick, was traveling. The two had also been shown the cafeteria, the library and the fitness center. Knowing they would be doing okay, didn't stop Patrick from worrying about the two females he found himself growing close to. He was the last to enter the lab.

"Sorry I'm late. I just wanted to check in on Miya and Lucy."

Kyle laughed. "Apparently Patrick found the only two good things in Province."

Ignoring Kyle, Patrick slid into his chair. As he did, Joshua stood from his own seat.

"I'm glad they're settling in. It's time for

you four to get back to work though. The other teams have already been to their second world."

Jess frowned. "Did they have any better luck than we did?"

The head of sandy hair bobbed up and down. "The Gamma team traveled to a place I think has great possibilities. It's called Fortine. The only drawback is that there would only be room for half our population. The Chamber of Guardians is setting up meetings with the leaders from that world. If we can find another Parallel World or two like that, we can give the people of Terrene some kind of option for where they would like to go when the time comes."

Now, Baileys' forehead creased. "What about the Beta team?"

Joshua shook his head. "Not yet, but their next destination is to a world called Gallatin, and it is showing a lot of promise."

Kyle smiled. "Hey maybe we'll get lucky too. When do we head to our new destination?"

Joshua smiled. "This morning, if you feel you're ready."

Jess frowned. "I filled Bailey in. Remember, you sent her shopping."

Joshua glared at Jess. "That was three days ago. I hope you did more than just fill her in."

Bailey shook her head. "Don't even listen to him Joshua. We've all studied the research you gave us. I like the place. It looks so peaceful. I love all those big trees around the Serenity Center. The name sure seems to fit what we've been seeing."

Joshua grinned. "I guess the four of you will be finding out. Do you have your EDAR Units?" When all four nodded, Joshua continued. "Good, I'm sending you all in together for this trip. After the last one, I've come to the conclusion I better keep you traveling as a group. We only have the one camera set up. I'd like you to get into that building and gather information. If you get the opportunity to investigate more of the area, go ahead and do it."

Joshua smiled. "That is, if you can stay out of trouble while you're doing that. Your track record hasn't been looking all that good lately."

Joshua took the time to look at each member of his Alpha Team before he smiled.

"You four are a hell of a team, despite the setbacks. I know you can't always control the circumstances that arise, especially now that we don't know as much as we usually do about these Parallel Worlds. That's my fault for trying to get in as many expeditions as possible."

Bailey shook her head. "Not your fault Joshua. I'd say an asteroid is most definitely in the category of a circumstance beyond all of our control."

Grateful for the acknowledgment, Joshua smiled. "Let's get the doorway open then and see if we can do something to fix that."

The crew headed to the doorway, while Joshua, at the computer, activated the system. He then turned to watch as his team stepped into Scintilla and the door shut behind them. Turning back to the screen, he watched the four cross the lush green lawn and make their way to the Serenity Center.

*   *   *

As they walked, Bailey frowned. "I can't figure it out, but there's something familiar about this place. I know I've never been here before, well, except on the computer screen, but…"
Bailey shrugged. "I don't know how to explain it."

Jess smiled. "Maybe you had a dream about it."

Kyle nodded. "Jess is right, this place does look like something that would come straight out of your dreams. Let's go inside and see if the interior is as special as the exterior."

The group seemed to take a collective breath and then walked toward the building. As they stepped on the front porch, that ran the entire length of the building, Bailey gazed in astonishment at the array of potted plants. Every corner of the porch seemed to be filled with blooming plants. Most of which, Bailey was sure she'd never seen before and would be hard put to give names to.
The door they stood in front of was almost twice as wide as a standard door. The ornate trim, a light peach color, stood out against the brilliant white that covered the rest of the building.

Kyle was the first to step ahead of the others to try the door, which opened easily. As Kyle pushed it open, the group stepped through into a large room where the potted plant theme was duplicated.

Seeing a woman, behind a twelve foot long, chest high desk that looked like the check in counter of a hotel, the team stepped over. The woman behind the counter smiled, but the crew saw curiosity in the wide set hazel eyes. "Welcome to Serenity Center, I'm Amira, what can I help you folks with today?"

Waiting for an answer, Amira brushed back a silver curl, which had come loose from the bun she wore, to fall down over her forehead.
Bailey smiled. "To tell you the truth, we were hiking and got lost. We saw your building and were hoping we could find some help here."

Now, the hazel eyes frowned and narrowed with suspicion. "Why would you be hiking out here? All the property in this area is private and we don't care for trespassers. This land is owned by the Serenity Center."

Jess stepped up to next Bailey. "I'm sorry, we weren't aware of that. We didn't see any signs up. This area is just so beautiful. I hope we're not in any kind of trouble."

The woman shook her head. "Of course not. What kind of help do you need? Do you want directions? If you're walking, I'm afraid the nearest town is at least ten miles from here."

Glancing around the room, Bailey then turned back to the woman, smiling broadly. "What is this place anyway? It's so majestic."

When Amiras' hazel eyes lit up, Bailey knew she'd said the right thing. "Like I said, this is the Serenity Center. We are a facility to heal physical and mental ailments. Our center, although large, only has a ten patient capacity. We believe the personal approach is the best way to help our patients. The residents here are mostly long term."

Both Kyle and Patrick, who had been standing behind Jess and Bailey moved closer. Patrick let Kyle do the talking. For him, like he'd stated before in the lab, there was something unnerving about the too perfect look of the hospital.

Putting on his most pleasant smile, Kyle stared at the woman. "Do you think it would be possible for us to have a look around?"

The look he got back was the exact opposite of the one he was trying to convey. "That would be impossible. No one but family can see our patients. Before that privilege is allowed to happen, they have to undergo a very strict screening practice."

The woman shook her head vehemently. "No, I'm afraid it would be better if the four of you left now. If you head back out that door, you'll see a sign at the edge of the lawn that points to the main road. It's only a half mile walk to the road. Maybe you can find someone there willing to drive you into the next town."

Glancing sideways, Jess could see the scowl on Kyles' face and took hold of his friends' arm. ""We'll do that. I'm sorry if we bothered you. Thank you for the directions."

Now all three of Jess' team mates were looking at him like he'd gone crazy. With a slight shrug, Jess gave the others a slight grin. "I think we better go. Sounds like it's quite a ways to the next town."

Turning back to Amira, Jess nodded.

"Thanks again, we'll get out of your way now."

As Jess herded the others out the front door, a crease of worry joined the other wrinkles on Amiras' face.

As soon as the four got outside and off of the porch, Kyle turned to Jess, his eyes filled with anger. "Why in the hell did you push us out of there Jess? We need to get some information."

Shaking his head, Jess stared at Kyle and then at the others. "In case none of you noticed, that woman wasn't giving out any details. From the way she talked, I don't think anyone else working there would be either. Let's take a look around the hospital grounds. Then, since we aren't making any progress going through the front door, I suggest we find a back way in."

The brown eyes that had just been glaring at Jess, now lit up as Kyle rubbed his hands together. "Oh, now you're speaking my language."

Staring at the two men, Bailey couldn't believe what she was hearing. "Maybe we should just forget about Serenity Center and try to get to the town that lady was talking about."

Patrick shook his head at Bailey. "I think we should find out what's really going on in this center. It doesn't look like a hospital to me and I thought that woman, Amira, acted really strange when Kyle asked if we could look around. There's something weird going on here. I can feel it and I also get the impression that this is the place we were meant to be."

Letting out a low moan, Bailey finally shrugged. "Looks like I'm outnumbered. Let's go have a look around, but keep a close eye out. That lady might have warned the others that work here about us."

Lifting his hand, Jess motioned toward an area just to the right of the doorway that had brought them from Terrene to Scintilla. "I think we better head toward that road Amira told us about. If anyone is watching, they'll think we're headed there in search of a ride."

As the others nodded their agreement, Jess led the way and the team started walking. As they hiked across the front lawn, Bailey stared in amazement at all the trees and bushes that were just off to the side and that seemed to stretch out as far as the eye could see. "This place really is gorgeous."

Smiling, Kyle nodded. "Not only that, but with all those trees hiding us, I think we should be able to make our way around to the back of that building without anyone spotting us."

Passing the sign that pointed toward the road the group slowly made their way to where they could see the main road. The one that would have led them to the town the woman in Serenity Center told them about. Instead of continuing on that path, the group, still following Jess, headed in a loop that brought them along the side and then to the back of the hospital building. Staying in the cover of the trees, the group was able to easily keep the large, exceedingly white, building in their sights. Seeing it without being seen, they group finally stopped.

Turning to the others, Kyle shook his head. "I don't know why we're being so cautious. I can't see a soul in sight anywhere near the building."

Jess shrugged. "That doesn't mean we're not being watched. They could easily have a security system set up. Right now, they could be taking a picture of that ugly mug of yours."

Running a hand through his brown hair, the dark eyes lit up. "Hey, people line up for the chance to get my picture. You must be thinking of someone else."

Patrick shook his head at the two men. "Personally, I don't find either of you that handsome."
Turning to Bailey, Patrick grinned, "No offense about your taste in men Bailey."

Laughing, Bailey shrugged. "What can I say, I can't help it, I love the guy. Besides, he has other talents. I think right now, we should be more worried about getting to that building and then inside. Anyone got some kind of plan?"

As the four stared at the building Bailey had just mentioned, Patrick sighed. "I think one of us should just walk up there. If nothing happens, the rest of us can just head up there too."

A look of disbelief crossed Baileys' face. "And what if something happens?"

It was Kyle who laughed. "Then we run up there and kick some butt."

Bailey stared at Kyle before turning to look at the others. "I should be the one to go."

Jess turned quickly to stare at Bailey. "Why would you go? I think one of us would be a better choice. You're not very threatening."

Bailey nodded. "Exactly why it should be me. If I do get spotted, no one would see me as a danger."

Not liking it, but knowing she was right, Jess finally nodded. "Okay, but if you see anything that looks even remotely out of place, signal us, or better yet, just high tail it back here."

Standing on the tips of her toes, Bailey kissed Jess on the cheek. "Don't worry, the last thing I want to be is a hero."
Bailey turned to look at Patrick and Kyle before she spun back to face the back of the building and she started walking.
As she took each step, she glanced warily from side to side, still wondering at the absence of people. It was a warm day with a bright sun shining. Even in a regular hospital they took people out to get some fresh air.
As Bailey got closer to the building, she saw not only a back door, but under a small awning, also a side door. Hoping that would be a better option, she headed for it, grateful for a large tree that partially blocked the view of anyone who might see her headed toward it.

That was also the reason none of the team members had noticed the door when they'd made their stealthy walk around the building.

Back in the cover of trees, three pairs of eyes were glued on the woman who looked even tinier than usual as she approached the large building. Those same eyes narrowed with concern as they saw Bailey move, not to the back door, but to the side of the building. Jess groaned. "Where the hell is she going?"

Kyle touched Jess' arm. "It'll be okay, Bailey knows what she is doing."

Just then, Bailey took a couple of steps toward the men and motioned for them to join her. All three sprinted across the back lawn and headed for Bailey. As soon as they reached her, Bailey pointed toward the building. "There's a side door and it's open. I looked in the window and could see a set of stairs going up and another set headed down."

All members of the Alpha Team hurried over to the door. Kyle was the first to reach it and turned the knob. Pushing the door in, Kyle took a timid step and then stopped as he listened for any noises. When he didn't hear anything, he moved away from the doorway into the small landing that sat between the sets of stairs so the others could crowd in, closing the door behind them.

Kyle pointed toward the set of stairs leading upward and whispered. "Let's try up there first."

Instead of answering, the others just bobbed their heads in acknowledgment. With Kyle leading the way and Patrick taking up the rear, the four slowly and quietly crept up the stairway. At the top, another landing led to a long hallway of closed doors.
Bailey frowned before speaking in a hushed voice. "Do you think those are the patients' rooms?"

Lifting his shoulders, Kyles' head tipped to one side. "I don't know, but I guess we'll find out."
Taking the hall, Kyle walked to the end and stood at the door that was closest to the far wall. He looked at the door, hoping for at least a number or more hopefully, a name. Seeing neither, he turned back to the rest of the crew. "Should I knock, or just go in?"

Patrick was the first to answer. "Just go in, you can always say you were lost."

Grabbing the knob, Kyle turned it slowly and opened the door. Leaning in, he saw an empty bed. By the looks of it, the occupant hadn't left too long before. Kyle stepped in with the others following close behind. The room looked like any other hospital room. It contained a twin size bed, a night stand and a small dresser. Another door was open at the other side of the room. Kyle went over quickly and leaned inside, glancing around to assure it was also empty. Seeing a vacant bathroom, Kyle blew out a relieved breath. Glancing over at the others, he grinned. "No one's in their either."

As Kyle walked back to the others, he saw Patrick staring at the blank wall directly across from the bed and frowning. "What's wrong?"

Instead of answering, Patrick pulled his EDAR Unit from his back pocket. Turning it on, he tapped the controls on the front and then pointed it toward the wall. "There's a doorway here."

Bailey, Jess and Kyle hurried over to stand close enough to Patrick to be able to look at the EDAR Unit. Looking up from the unit, Baileys' blue eyes widened. "How did you know?"

Patrick shrugged. "I thought the wall looked like it was moving slightly. At first, I thought it was my eyes playing tricks. That's why I brought out the machine."

Jess sighed. "Thank goodness Joshua got those for us. What do you think it means and where do you think the damn thing goes?"

Patrick shrugged. "I'm not sure, and without a computer to research it, I don't think I want to find out."

Kyle nodded. "That's for sure. The EDAR can read the atmosphere, but by the time we stepped through and checked, we could all be dead. No thanks."

Jess was looking curiously at the wall. "Why would a hospital room have a doorway like that? They always lead to Parallel Worlds, or at least all the ones we've found have."

Bailey frowned. "Maybe the Serenity Center is more than what we were told it is."

Kyle grimaced. "I don't like being lied to. Let's go check another room."

Leaving the empty room, the four walked down the hall to the next closed door. This time it was Patrick who opened the door with one hand, holding his EDAR Unit ready in the other.

When the group stepped in they stared in surprise at a young woman sitting up on the bed. Large green eyes stared at the intruders as they all moved into the room and shut the door behind them. As Bailey looked back at the face, she frowned. Just like she had felt when she'd first seen the building they had snuck into, there was something familiar about the woman. Looking from the womans' face to her hands, Bailey gasped. Each of the tips of the womans' fingers were bandaged.

Beside Bailey, Patrick turned to the wall that was opposite the womans' bed. He held up the EDAR Unit, even though he already knew what the result would be. Catching the motion from the corner of her eye, Bailey turned from the woman to look at the wall. A small cry escaped her throat before she could stop it. The wall was covered with scratch marks and Bailey knew now why the bandages covered the womans' fingers. Turning away from the damaged wall, Bailey stepped toward the bed, her eyes filled with sympathy. "Don't be afraid, we won't hurt you. My name is Bailey. What's yours?"

The green eyes grew larger and glanced nervously back and forth between the four strangers before settling back on Bailey. "My name is Gina. Do I know you?"

Lifting her eyebrows in surprise, Bailey then smiled. "I'm not sure, but for some reason, I feel like you might."
Then she frowned. "Although, I don't know how that could possibly be true."
Pointing at Ginas' fingers, Bailey winced. "Are you okay?"

Embarrassed, Gina slid her hands under her legs. "I'm fine."

Trying to keep a smile on her face, Bailey took another step closer. "I didn't mean to…I mean it's none of my business." Bailey shrugged then and stopped talking, knowing her explanation was only making Gina more uncomfortable.

Behind Bailey, the three men were all standing by the wall, looking at the area that looked like someone had tried to claw their way through.

Bailey had moved even closer to Gina. Now the woman had a slight smile on her face. "You're from there aren't you? From the other side? I've seen your face in my dreams."

After a moment considering the question, Bailey nodded. "Is that what you've been doing, trying to get to the other side?"

Gina nodded. "We all have. That's why we're here."

Bailey frowned. "Who put you in this place?"

By now the men realized Bailey and the woman were having a conversation and had turned around to listen. Seeing Ginas' gaze move in the direction of the men, Bailey smiled. "It's okay, they're friends. They know about the other side too."

Gina sighed, then shook her head. "It doesn't really matter. I can't get through anyway."
Now, Gina held up her hands for all to see. "I've tried and tried."

Baileys' face filled with anger. "Who put you here and made you do that?"

Gina shook her head. "It's not like that. I wanted to come here. This is a research center for people like me. The ones who can feel the other side."

Now all four members of the Alpha Team stared wide eyed at Gina. Jess stepped closer. "You can feel it?"

Gina nodded. "In my dreams I see the worlds, the people too."
She pointed at Bailey. "I've seen you."

Gina frowned. "Them too, I think. I don't always remember who or what I see. When I wake up, the dreams fade fast, but that's when I find myself going to that wall and trying to get through and why…"
Gina wiggled her fingers. "…why this happens."

Stepping up beside Bailey, Patrick looked at Gina. "Do you have a main doctor here who helps you?"

Gina nodded. "Sure, Doctor Quentin. In fact, he'll be coming by soon. He usually comes just before lunch to talk with me about the dreams I have. In the afternoon, I see Doctor Messer. She's a therapist."
Gina laughed. "I call her my dream coach, why?"

Patrick smiled. "I think we may be able to help you and then find these other worlds you are feeling. Do you have computers here?"

Gina nodded. "Sure, there's a computer lab in the basement. Actually, none of the patients go down there, but I know about it."

Gina frowned. "I don't understand, why are you asking about the computers?"

Now, Patrick laughed. "I know someone who may be able to help by setting up a computer program that might open the door to your other worlds."

The green eyes lit up, but then changed as Gina frowned. "Some of the other worlds aren't good places."

Patrick nodded. "I think we can help with that too. Would it be okay if we waited in here until the doctor comes?"

Turning to Bailey who nodded her okay, Gina sighed. "I guess it will be okay. I have to warn you though, we are hardly ever allowed visitors. Once a month family gets to come by, but even then we only get short visits. We aren't allowed to talk about our treatments. If I wouldn't have seen Bailey in my dreams I wouldn't have said anything to any of you at all. It was just such a shock." Gina glanced at Bailey. "A nice shock though. In my dreams you're in a good world."

Bailey smiled. "I suppose since you're letting us stay here, I should officially introduce my friends. The one with the big ideas is Patrick."

Turning, Bailey grabbed Jess' arm. "This is Jess and the other guy standing there is Kyle. We're in fact more than friends, we're team mates. I know this must all be confusing and sudden. I'm so glad you're handling the bombshell we've dropped on you so well. I don't think I'd be as gracious in the same situation."

Gina shrugged. "Like I said, I've seen you plenty of times. My dreams are very vivid and real to me. Sometimes it's hard for me to separate my waking world from the ones I walk through in my dreams."

As Gina said that, tears filled her eyes. Seeing the others worried looks, she ran her fingers through her short reddish brown hair and shook her head. "I'm not crying, I'm relieved. You don't know what it's been like. My family and friends thought I was losing my mind. In fact, when you people first stepped in my room and I saw Bailey, I thought my mind had finally snapped. It was like my dreams had found their way into my waking world. When I heard Bailey talk, I realized that hadn't happened. The relief of knowing that overrode any fear I felt."

Hearing that, Bailey felt like crying in empathy for the woman on the bed. Moving closer, she reached out and touched Ginas' hand, careful to avoid the bandaged tips. "I'm sorry to put you through that and sorry that you've had to deal with all of this. I think we can help you discover what makes you feel the other worlds. You must be very special to be able to do that. I've been to some of these places, but I've never been able to just feel them."

Gina barked out a harsh laugh. "I don't feel special. Most of the time I'm just scared to death."

While Bailey continued to talk to Gina, the rest of the Alpha Team walked over to examine the marks Gina had left on the wall. Kyle reached down and ran a hand lightly over the scarred surface. He wondered which world was beyond that wall that would compel a person to try and crawl through. As he was about to ask Jess and Patrick what they thought, the sound of the door opening pushed the question from his mind.

Everyone in the room stared at the man standing in the doorway. The light blue eyes looked even wider behind the thick glasses the man wore than they were. "What's going on in here? Why are you people in this room? Who the hell are you and how did you get in here?"
The man looked over at Gina. "Are you okay?"
Gina had only given a half nod when the man began turning away. "I'll call security."

As Gina screamed "No", the man stopped and stared at her. Gina swallowed the nervous lump in her throat. "I told them to stay. They're here to help. Doctor Quentin, you need to hear what they have to say, please."

At the sound of desperation in Ginas' voice, the doctor ran a hand through the short white beard that matched the color of his thinning hair. The blue eyes narrowed. "Why would I listen to people who have broken into a private facility?"

Before Gina could answer, Patrick spoke. "Because we know about the other worlds."

Staring at him moment in stunned silence, the doctor then turned his gaze to Gina. "You told them about the other worlds?"

As Gina shook her head, Patrick took a step closer to the doctor. "Gina didn't have to tell us about them. We're from one of those other worlds. I can prove it to you, if you'll give me a minute. If you don't believe me when I'm done, you can go ahead and call your security."

Looking at Patrick, Doctor Quentin was now more curious than nervous. He slowly nodded. "Start explaining, but if I don't buy what you are saying, the four of you will be arrested and detained."

As Patrick nodded, the rest of the Alpha Team moved closer to him as a sign of solidarity. Pointing toward the marred wall, Patricks' mouth straightened in a grim line.

"There is a world beyond that wall. I don't know how Gina can feel it, but I do know how to open a doorway into it. The four of us are part of a program whose sole intent is to find and explore other worlds. In our World, we have computers that are programmed to search out those worlds." Patrick pointed at Gina. "It appears that you have the same thing in this special young woman that we found through technology." Patrick shook his head. "That is an incredible concept."

Patrick held up his EDAR Unit. "This unit can find the areas where electrons are diffused as the magnetic lines cross."

Behind the thick glasses, the blue eyes grew large. Doctor Quentin knew all about the magnetic lines. The Center sat on top of the spot where the magnetic lines converged. Encouraged by the look, Patrick continued. "I can open a doorway and show you. I won't, however, leave it open. Without my computer program, I have no idea which world it will open into, or what might be waiting on the other side."

The doctor nodded. He knew, by talking to Gina, not all of the worlds were places one would want to wander unknowingly into. "Okay, open the doorway. If you can do that, I'll forget all about calling security."

Nodding, Patrick pointed the EDAR Unit at the wall. All eyes in the room were glued on the spot. A glimpse of a landscape could be seen, although hazy and distorted, then the doorway closed. The wall, with its scratch marks, rematerialized.
Running a hand through his white hair, Doctor Quentins' stare went from face to face in amazement. The gaze stopped to rest on Ginas' smiling face. He smiled back, then turned to Patrick. "I believe I'm ready to talk now."

Chapter 8

Hours later, the Alpha team were all seated in Joshuas' office. After talking to Doctor Quentin in Ginas' room and a quick trip to the computer lab to see what was available for technology in this world, the team had promised they would bring Joshua back to talk to the doctor.

After the initial shock of finding four strangers in his patients' room, Ray Quentin had been delighted and extremely interested in what the Alpha Team had to say and anxiously awaited their return.

In Joshuas' office, Patrick had done most of the talking, as he explained what they'd found at The Serenity Center in Scintilla. Joshua was frowning, his green eyes darkened with concern.
"I don't quite understand what this Doctor Quentin was trying to accomplish. Did he think one day one of his patients would

actually claw their way through into another world?"

Patrick shook his head. "I don't think he was trying to do that. The doctor and his assistant were hoping to learn about the other worlds by examining what the patients saw in their dreams. Doctor Quentin's assistant is a psychiatrist."
Frowning, Patrick turned to Bailey. "What was that ladys' name?"

Bailey smiled. "Gina said her name was Doctor Messer. She called the doctor her dream coach. I'm sure this Doctor Messer didn't really believe in actual doorways either."

Joshua smiled. "I'll be glad to share the technology we have. I'm also hoping that we can work out an exchange."
Seeing the others' curious looks, Joshuas' smile broadened. "Our technology in exchange for a place for some of our people to go when the time comes. According to our research, they have the room and their world is similar to our own."
Joshua stared at Patrick. "Would you be

155

okay returning to Scintilla with me? I'd like the rest of the team to prepare for another journey. There's a world we've found and named Firmon that I'd like you to take a look at."

As Patrick started to nod, Kyle frowned. "What do you mean you named the world? Why doesn't this Firmon already have a name?"

The green eyes across from Kyle grew serious. "It's an empty world. No one lives there, at least no one we've seen. Firmon isn't the type of place a lot of people would be willing to move to. It would be a new beginning and I mean that literally. The new occupants of Firmon would be building a new civilization from scratch."

Kyle grunted. "I wonder if we could do that without screwing it up."

As Joshua shook his head, he let out a sigh. "That I don't know, but I know a few people who would like to give it a shot. So for now, the Gamma team found Fortine. The Beta Team ensured Gallatin will let some of the population in. If we can work out a deal with Scintilla, that gives us three destinations. If we can build on Firmon and maybe find another world or two that would allow us to come in, I think we'll have a way to save all of Terrenes' population before that damn asteroid destroys our world."

Joshua again returned his attention to Patrick. "I'd like to leave for Scintilla the first thing in the morning."

Patrick nodded. "I'll be ready."

Josh looked at the rest of the group. "I'd like the rest of you to prepare for the trip to Firmon so you can leave in the morning also. I'll get one of the scientists to man the computer. The three of you need to head to the lab. The information, what little there is, has been added to the computer, we only have one camera set up, but we've been able to get quite a few still shots. At the very least you'll be able to get a feel for the place."

Patrick stood first, followed by the others. "I'd like to go and check on Lucy and Miya, then I think I'll get some rest. I'll be back at the lab first thing in the morning to start our traveling though."

Nodding, Joshua grinned. "I can't wait, I'm looking forward to seeing the Serenity Center and talking to the doctors. While Patrick and I head there, the rest of you can take that journey into Firmon."

Leaving the room, Patrick turned to look at his fellow crew members. "I guess I'll see you in the morning."

Kyle laughed. "Tell Lucy and Miya we said hi and don't stay up too late. Remember, you'll be traveling with the boss."

Patrick laughed. "Don't remind me. At least we managed to stay out of trouble on our last trip. That should help ensure Joshua's in a good mood. I'll see you in the morning."

Patrick watched his three co-workers walk toward the lab before turning and heading to Lucy and Miyas' apartment.
When he reached their door, Patrick hesitated a moment before knocking. He felt something special towards both Lucy and

Miya, but wasn't sure how Lucy felt. He knew that he'd made an exceptional connection with Miya. He hoped he'd made one with Lucy too.

Drawing a breath, Patrick lifted a hand and knocked. It was Miya who answered the door. The five year old was dressed in a bright yellow outfit today and to Patrick looked like a sunshine filled day.
"Patrick, come in and see what I been doing."

Smiling at the exuberance in her voice, Patrick stepped in. Lucy looked over at him from where she stood behind the counter that divided the kitchen and living room. "Patrick, how nice to see you. Did your trip go well?"

Patrick nodded. "I think so."

Before he could say more, he felt Miya tugging at his hand. "C'mon, you gotta see."

Lucy shook her head. "Sorry, she's fascinated with the computer."

Patrick laughed. "That's okay, I'm the one that brought it here to begin with. Let me see what Miya's doing and then I can fill you in on my trip."

Lucy nodded. "Would you like some coffee?"

As Miya pulled him away, Patrick nodded. "That would be great."

Following Miya, Patrick went to the far end of the living room where he'd set up the computer on a small desk for both Lucy and Miya to use. He was guessing that Miya had gotten the most use out of it. He listened patiently as the preschooler told him all about the great games she had found to play. Patrick was pleased to see that most of the games, although fun for Miya, were also learning games. He could easily see, that at the age of five, Miya had already learned more than a lot of people he knew five and ten times her age.

When Miya finally finished sharing, she went back to the computer and Patrick joined Lucy in the kitchen.

The two sat at a small table with their coffee in front of them. Lucy was smiling. "Miya just loves that computer. I'm afraid she's figuring it out faster than I am."

Nodding, Patrick laughed. "It always amazes me how quickly kids learn new things. Don't worry, you'll catch up. It's still hard for me to think that Province didn't have any computer technology. All the worlds are linked and I just thought something like that would carry over to the next world. I should have known better, some of the worlds we see are completely different than this one."

Lucy laughed harshly. "Province is one of those that is worlds apart from here. Not just in knowledge either. Next to this place, I think it must have looked pretty bad."

Patrick smiled. "I hope that means you like it here."

Nodding, Lucy's eyes grew wide. "I love it and so does Miya. Even better, Joshua said he's going to give me a job. I'm going to be helping out in the infirmary. At first I'll just be cleaning up, watching the front desk and assisting where I can, but Joshua said I could get training as a nurse too."

Taking a sip of his coffee, Patrick studied the lovely face he saw over the top of his mug. Setting the cup down he smiled. "I'm so glad things are working out. You and Miya were really thrust into a crazy situation."

Shrugging her shoulders, Lucy smiled. "I can't complain. Never mind about me though, I want to hear about your trip. I mean if it's not a secret. I wouldn't want to get you in trouble."

As Patricks' shoulders lifted in a shrug, he laughed. "If it was a secret, it wouldn't be for long. The people of Terrene have to be told. They'll be moving into the places we find soon."

Lucy frowned, her eyes sad. "Such a tragedy. Are you positive about the asteroid?"

Patrick nodded. "There's no doubt. This world is definitely in for a cataclysmic clash with that thing. We've already been able to get a couple of worlds lined up for the evacuation though. There's also a good possibility that Scintilla will be added to the list. I'll be going back in the morning with Joshua."

Lucy stared at Patrick and then listed with rapt attention as Patrick told her about first his journey to Scintilla and then the other places he traveled to. Before either realized it, time had slipped away. Patrick accepted Lucy's invitation to dinner and then he spent time with mother and daughter until it was Miyas' bedtime.
At the childs' request, Patrick tucked her into bed before joining Lucy on the living room couch. He was surprised and pleased when Lucy took his hand in hers. "Thanks for being so patient with Miya. She already thinks the world of you."

Smiling, Patrick stared into the blue eyes. "I feel the same about her. What about you, what do you think about me barging in and taking up your time?"

A shy smile appeared on Lucys' face. "I think that is something I could take a lot more of."

Relief washed over Patrick as Lucys' words touched his heart. Leaning over, he pulled her to him. The kiss was longer and much more than either of them expected it to be. Patrick reluctantly pulled away.
"If I'm going to travel with Joshua in the morning, I think I better get out of here. Much more of that and you'll never get rid of me."

Lucy nodded. "I'll let you go tonight, but next time, there's no guarantees."

As he stood, Patrick realized that was something he couldn't wait to find out.

After Patrick left, Lucy sat on the couch thinking of how much her life had changed in just a few short days. Coming to Terrene was like a dream come true. She only hoped that wherever she and Lucy had to go before the asteroid hit, that place would include having Patrick at her side.

*   *   *

As Patrick entered his apartment, he couldn't believe how lonely the place seemed nor how alone he felt. The only people he spent any kind of time with were the others in the Alpha Team. Now that he'd met Lucy and Miya he realized just how empty his world had been.

As Patrick got in bed, he lay with a smile covering his face as the picture of two special people filled his mind. Then Patrick fell into the world of dreams.

Chapter 9

The next morning when the Alpha team entered the lab after a quick breakfast in the cafeteria, Joshua was waiting. He had another man with him. The two turned from the computer when the door opened. Joshua stood. "Finally, the Alpha Team makes an appearance."

Kyle looked at his watch. "I think it's still pretty damn early. I swear you must get up before the sun."

Joshua shrugged. "I'm sure you've figured out by now that I usually do."
He turned to point at the man still seated in front of the computer. "This is Cade Thompson, he'll be the one monitoring our trips today."
Joshua let everyone exchange 'hellos' before he continued. "Patrick and I will make the first trip. After we're gone, the rest of you can head out to have a look at

Firmon. We can all meet later to discuss what we find."

Getting nods of agreements from the Alpha Team, Joshua turned to look at Cade. "I think we're ready. Do you want to set up for the Scintilla trip?"

Nodding, Cade turned to the computer while Joshua and Patrick stepped to the doorway. As soon as the doorway opened, the two walked through and headed for the Serenity Center in Scintilla.

The remaining three members of the Alpha Team waited the few moments it took for Cade to adjust the computer for their own trip. When he finished, they made their way to stand in front of the doorway. When it opened again they stepped through into the World known as Firmon.

As they crossed over and the doorway shut behind them, the three stared in wonder at the empty world that sat in front of them. Pointing ahead of them, a grin covered Baileys' face. "Oh my goodness, would you look at this place. The computer video doesn't even come close to doing this place justice. Have you ever seen anything so beautiful?"

Kyle was nodding. "I thought Scintilla was a paradise, but this place definitely has it beat."

As they took their first timid steps in the lush, green grass, Jess reached out a hand to touch one of the many bushes that surrounded them. He reached down and pulled a berry from the bush. He held up the violet colored fruit. "I hope these are good to eat. There must be hundreds of these bushes right here."
Pulling a small container from his pocket, Jess opened it and dropped the berry in before replacing the lid. "We can have that analyzed back at the facility. Let's go see what else this place has to offer."

As Jess started to walk away, Kyle grabbed his arm. When Jess turned back to look curiously at his friend, Kyle handed him a small rod. Jess frowned. "You brought stunners?"

Kyle nodded. "We don't know anything about this place. I thought we could use some kind of protection."

Bailey was frowning. "You know how Joshua feels about us having weapons."

Kyle nodded. "I do, but it was a stunner that saved Jess' ass when that regulator back in Province had him down on the ground. Listen, you know as well as I do the stunners aren't for killing someone. I grabbed these for our protection. If we get in a predicament these could buy us the time we need to get out of it. Until we learn more about this place, I feel better having one of these close by."

Putting the stunner in his pocket, Jess nodded. "You're right, we don't know what we might find in this place. C'mon let's get moving."
Frowning, Jess pulled out his EDAR Unit. "Does this thing have a compass or something on it?"

Bailey nodded. "Here, give it to me. Have you ever even really looked at the unit Jess?

It can do all kinds of things."

Rolling her eyes at Jess, Bailey found the right screen and handed the EDAR back. "Here, I think you can figure it out now."

Giving her a disgusted look, Jess took the EDAR, then after looking at the screen, pointed to the left. "Let's have a look over that way first."

The three only hiked a short distance when they hit a large stand of trees so thick they couldn't see through to the other side. Kyle moved ahead of the other two and began pushing branches aside and making his way through the dense green foliage. Following behind Kyle, Jess and Bailey tried to avoid the branches whose momentum swung them back at the pair.

A few moments later they heard the sound of Kyle letting out a long, low whistle. "Will you look at that?"

Jess and Bailey stepped up beside Kyle and then all three stared at a large lake filled with turquoise colored water. On the other side of the lake the large hills were reflected in the waters' mirror surface.

Bailey pulled out a container from her pocket and carefully took a sample, avoiding getting any of the water on her bare skin. "I hope this is drinkable."

Jess pointed toward the mountains on the far side of the lake. "I think we might just be in luck. That looks a lot like snow to me on the tops of those peaks. It probably melts and flows down to form this lake."

Kyle was nodding. "We need to find a good place or two to grab some soil samples. If this is a good supply of water, we could easily grow our own food here."

Jess nodded. "That would take care of the basics, air, water and food, but we wouldn't have much else on our side, no power, no housing…"

Bailey interrupted him. "No pollution, no crowds and a way to begin fresh. I think it sounds perfect, I like it."

Jess laughed. "I don't think you realize just how hard that would be."

Bailey shrugged. "I think it would be worth a try."

Kyle nodded. "And there's always a doorway if you couldn't make it and had to leave."

Holding up his hands, Jess smiled. "I give, this place might have possibilities after all. C'mon let's go check around the other side of the lake."

It took about a half an hour for the trio to walk to the far side of the lake following the shore line. Reaching the area, the team found more bushes, trees and to Baileys' pleasure a collection of colorful flowers that gave off an aroma like sweet perfume. As Bailey bent down to get a better look, she saw something moving in the bushes just ahead of her. Frowning, Bailey straightened up and pointed. "I don't think this place is quite empty."

Jess and Kyle turned to look where Bailey was pointing. As the three watched, an animal, a bit bigger than a squirrel, stared back at them. It stood on its' hind legs and held a piece of fruit in its' front paws. Not only was the animal the size of a squirrel, it was also similar in shape. Bailey smiled. "Isn't it cute?"

Kyles' eyes narrowed. "Cute doesn't mean safe."

As soon as the animal heard the sound of the crews' voices it turned and scampered away, running into a small hole at the base of the mountain. Bailey laughed. "I think it went home."
She turned to Jess. "Should we go look in that cave?"

Jess shook his head. "I think we should head back. We know what this area of Firmon has to offer. I'd like to see if we can open a doorway to other parts of this world and find out if all of Firmon is a paradise."

Bailey frowned. "Even if it isn't, there's enough room here for a good size town. Quite a few people would be able to build homes, plant food and live a good life here."

Jess shrugged. "We'll have to check out the samples we got first."

Kyle nodded. "That reminds me, I better get those soil samples."
Kneeling, Kyle took a sample, then he stood again. "I'll grab another one back by the doorway." He stared at Jess and Bailey. "Are you two ready to head back?"

The two nodded and the team made their way back toward the doorway. When they got close, Kyle got another soil sample and then Jess pointed his EDAR Unit toward the area close to where he knew the doorway was. If he would have taken time to study the area he would have seen the slight shimmer in the air the doorways cause, but using the Unit was much easier.
A moment later, a green light on the unit grew bright and Jess smiled. "There's the doorway. Is everyone ready to head back?"

Kyle rubbed his stomach. "I am, it must be lunch time by now."

Using the unit, Jess opened the doorway, thinking how much better this was than waiting for the operator on the other side to open it for them.

As the three entered the lab, Cade, who had been watching them approach on the monitor, turned to look at the trio. "You're back first, how did it go?"

Jess smiled. "I think it went well. We brought back a few samples to have analyzed. If they check out we can add Firmon to the list of worlds we can travel to."

Kyle laughed. "That is if the people headed there don't mind roughing it. Going to Firmon is like taking a rugged camping trip. It does have possibilities though."

Bailey looked at Cade. "You said we were the first back, haven't Patrick and Joshua returned yet?"

When Cade shook his head, Bailey frowned. "I thought they'd be back before us."

Jess shrugged. "We weren't gone all that long. Let's take those samples to get looked at. I'm sure they'll be back soon."

Kyle nodded. "Jess is right, besides I'm starving. We can grab some lunch after we drop off the samples and then we can head back here."

Leaving Cade to watch for Patrick and Joshua, the three headed to drop off their samples before walking to the cafeteria. Getting their food, the three found a table and sat down. Kyle and Jess ate heartily, but Bailey, still excited about Firmon, only picked at her food. Looking at her, Jess frowned. "What's the matter Bailey? You're awfully quiet and you've hardly touched your food."

Bailey twisted her lips in contemplation. "I just can't stop thinking about Firmon. It's just such a perfect world."

Staring at her, Kyle shrugged and then scowled. "A perfect world doesn't exist, at least I've never seen one."

Opening her eyes wide, Bailey stared in shock. "Are you kidding me? I think it can. You're just too much of a pessimist Kyle."

Taking a bite of his sandwich, Kyle shook his head. He chewed hastily and then swallowed so he could answer. "I think I have good reason to feel that way. After the worlds we've seen, how can you be optimistic? People pollute their worlds, or try to make themselves master over the populations. If you ever were able to find a perfect world, someone would come along and try to screw it up."

Bailey sighed. "I think we could do it. You just need the right people to set the ground work."

Pointing a finger at Bailey, Kyle moved it back and forth, then tried to give her a smile. "You are looking at how you'd want this world to be, but believe me, there's always someone who won't like what you're doing."

Not liking what she was hearing from Kyle, Bailey turned to Jess. "Now you're the one who is being awfully quiet. What do you think? Is Kyle right?"

Not answering for a moment, Jess studied Baileys' anxious face. He wanted to reassure her that the hopes she were holding on to were right, but wasn't sure if he could. Finally, he drew in a breath and then released it with a sigh. "I think you're both partially right. Although I don't think there's such a thing as a perfect world or a perfect person, I do believe you can make a world that's better than the ones we've seen." The brows above the dark eyes drew down in a frown. "I think you're forgetting one thing Bailey. You can't select the people you want to start a new world. The reason we're trying to find several worlds is not only to find room for everyone, but we are also giving them a choice of which world they want to live in. If you closed off the doorway to those you felt didn't fit into your perfect world, then you'd be making yourself master over that world. You wouldn't be any better than those you were keeping out."

Now it was Bailey who had a frown on her face. "I guess I never thought of it that way."

Jess nodded. "That doesn't mean you still couldn't work toward building the kind of world you want. You're thinking about starting over in Firmon. Have you given any thought to all that would entail? Not a lot of people, having gotten used to life here, want to go without power, running water or heading to the grocery store for their food."

Bailey nodded. "Maybe I'm hoping that less people will want to settle in Firmon. I'm sure there will be some who would prefer it though. I for one would love to grow my own food and live without what people think are the luxuries of life."

Jess stared at her. "And what if you do enter the doorway, start a new life, and then it doesn't work out? What if when that asteroid hits Terrene, all doorways close? I know earlier Kyle said you could come back if things didn't work out, but the fact is, there's a high possibility that the doors closing is exactly what will happen. Joshua will be taking his technology to Scintilla, so they might have some doorways there, but Firmon is one world that doesn't have any technology at all and cancels out that option."

It only took Bailey a split second to grin and come back with an answer. "Maybe that would be for the best. If we don't have the option of leaving through a doorway, we'd try harder to make things work."

Looking at Jess, Bailey could read the admiration in his eyes. But when Kyle laughed, she turned to him, and was surprised to see that same look echoed in his eyes. "Remind me next time I have to argue with someone to bring you with me. You make a damn convincing case. Maybe I'll have to rethink a life on Firmon myself."

Bailey smiled. "I'd be honored to have you there. In fact, I was kind of hoping the whole Alpha Team would end up together in the same world."

Kyle sighed. "We'd have to give up traveling if we choose Firmon. That's one thing to contemplate."

Standing, Jess shook his head. "None of it matters until we finish our research anyway. Let's head back to the lab and see if Joshua and Patrick made it back yet."

After clearing off their table, the three left the cafeteria and made their way back to the lab. As they stepped in, they were relieved to see Patrick standing behind Cade at the computer.

Bailey ran over and pulled Patrick into a hug, "What took you guys so long? We came back hours ago."

It was Joshua who answered. "You can blame me. I wanted to fill Doctor Quentin in on the science behind the doorways. I'm afraid once I start talking about all that, I get carried away. I have to tell everyone though, I am really optimistic about rebuilding the Postern Project next to or even in the Serenity Center. It might take a month or two, but I think we can make a smooth transition upgrading their technology to match our own."

Joshua looked at each of the team members.

"How did your journey go? What did you think of Firmon?"

As Baileys' eyes lit up, she did the answering. "It's beautiful Joshua. As close to a natural paradise as you could get."

Jess shook his head. "I don't think we've seen enough of that world to be sure. I'd like to open other doorways into Firmon if we can. The section we saw was all that Bailey said, but I'd like to travel to the other side of that world to see what it is like. Here in Terrene and in most worlds we travel to, there's technology that connects all parts of the world and we don't have a lot of surprises, but in Firmon we don't have a way to find out except by exploring."

Joshua nodded. "That's a good idea and one I'm sure we can look into. I'll get the researchers busy on that. I also need to check in with the Beta and Gamma teams. They each went to explore a new world. I'm excited that our options are growing. There's another trip I'll have to make and soon."

The team all looked curiously at Joshua and he sighed. "I need to make a trip to see the Chamber of Guardians. Now, with the option of a few worlds available, I think they will be ready to go public with the information of what lies in Terrenes' future."

The crew was silent a moment, thinking about what a tremendously hard job they all had ahead of them. Kyle was the first to shake his head. "I don't envy the guardians that job. I just hope we don't have a panic on our hands."

Looking at Kyle, Joshuas' head bobbed in agreement. "It won't be easy, but hopefully knowing there are alternative places to go will calm people fears."

Beside Joshua, Patrick grinned. "I think knowing Joshua will be moving his own project to Scintilla will help with that too. When the Chamber of Guardians makes their announcement to the world, they'll be able to include what Joshua has been doing here. Knowing he's been working to save our world and will continue that work in

Scintilla should go a long way in easing their worries."

Patrick yawned. "Sorry, but if I'm not needed, I'd like to go and grab a shower. I'm also looking forward to a good nights' sleep."

Placing a hand on Patricks' shoulder, Joshua nodded. "Thanks for all your help. Go ahead and take off." Joshua turned to the others. "In fact, you should all take a couple days off. I'll get that research started on other doorways into Firmon. Your team can get back together then and do some exploring. I'd say, for now, you have all earned a couple days rest."

Excited for the days off, the Alpha group left the lab and then went their separate ways. Kyle headed to the facilities one bar. He felt the need to spend time with the people he didn't know how long he'd be able to pass time with. Of all the Alpha team members Kyle was probably the only one that didn't mind being around crowds. Not all the time, but once in a while he craved the camaraderie of a large group. Patrick was probably the one in the group the most opposite of that. He liked to be alone, that was until just recently. Right now, all he wanted was to grab a shower before he could head over to see Lucy and Miya. The last two members of the Alpha Team headed to their apartment for two days of spending some much needed time as a couple.

In their apartment, Bailey had just closed the oven, where she'd put their dinner to cook. Jess was seated at the kitchen table. Turning away from the oven, she smiled at him. "That'll take about a half an hour. Anything special you'd like to do to fill the time?''

Nodding, the brown eyes gleamed. "Oh, I can think of something, but I think I'll hold that thought until later when we have more time."

Looking at him, the blue eyes danced. "I guess I can wait, just remember, I'm holding you to it."
Moving to the table, Bailey sat down opposite Jess. "I want to talk to you anyway."

185

Staring at the woman who had been his partner in traveling for five years and in his bedroom for three, Jess smiled. He already knew what this talk was going to be about. Instead of sharing that information with Bailey, Jess only nodded. "What should we talk about?"

Bailey blew out a breath. "How about a million things? So much has happened lately."

A mischievous look appeared in Jess' eyes. "That's true, but I have an idea you have one thing in the front of your mind."

Bailey laughed. "You really think you have me all figured out, don't you?"

Jess shook his head. "I wouldn't even dare say that, but I have a feeling you've had Firmon on your mind since the doorway opened."

The head of brown hair nodded slowly as Bailey gathered her thoughts. "You're right, I loved the place the moment I saw it. If I asked you, would you join me there Jess?"

Reaching across the table, Jess took Baileys' hand. "I'd join you anywhere Bailey. You should know that. You're the most important person in my life and I love you. Are you positive about Firmon though? Have you even thought about what going there will entail? Or what all you would be giving up?"

Shrugging, Bailey gazed into Jess' brown eyes. "I think I have, maybe even before I ever laid eyes on Firmon. I guess somewhere in the back of my mind, I've always thought about what that would be like. Just imagine, to start a new world." The blue eyes lit up.
"Despite Kyles' pessimistic objections, I think the right people could lay the foundation for, if not a perfect world, one damn close."

Lifting Baileys' hand, Jess brought it to his lips. "If the rest of Firmon is even close to what we already saw and if it would make you happy, then yes, I would give up everything to start over with you in Firmon."

Bailey felt like crying with relief and happiness. "No wonder I love you." A smile flitted across Baileys' face.

Jess nodded at what she was saying. "I don't think we'd have to give up everything. If Joshua can move the Postern Project from Terrene to Scintilla, I think we can manage to move a few things from home through the doorway and into Firmon."

Bailey smiled. "Sounds like a plan to me. Maybe I should get our dinner out of the oven. We can eat and then see if we can find a way to seal this deal."

Jess laughed. "I think I have an idea for the perfect way to do that."

Chapter 10

The Alpha team actually received three days off before they found themselves back in the lab with Joshua manning the computer. The four friends took seats beside Joshua, so they could also look at the computer screen. Joshua turned off the screen, "I don't think I'll show you what we found. I think it would be better to surprise you. I can tell you from what we saw, these new places are as safe as the one you entered into the other day. We were able to set up two new doorways into Firmon."

Josh nodded. "I can't wait to see what we'll find then. I'm glad you set up two locations, it's so different heading into a place no one inhabits."

Joshua laughed. "While the four of you were lounging around, some of us were busy working. The other teams have each found one more world to add to the list of destinations. Now we have five alternate worlds, six if we can count Firmon. I think we are really going to be able to save the people of Terrene."

Kyle was frowning. "Have you give any thought to the possibility that some people just won't leave, no matter what the future is for this world?"

The green eyes saddened. "Unfortunately I have. I can't force someone to leave their home. I'm afraid that not only will some not leave, but we might see quite a few finding a way out long before that asteroid hits."

Bailey frowned. "What does that mean? How could they find a way out?"
Then the blue eyes enlarged as Bailey realized what Joshua was saying. "Do you mean suicide? Why? I mean we are setting up places for them to move to."

Jess touched Baileys' shoulder. "I'm afraid we'll find some people who can't stand the idea of being refugees in a strange world and will either end their lives before the asteroid hits or stay here and let that asteroid end their lives for them."

Shaking her head, Bailey had to admit the possibility was there, even though she hated that it was. "I never even thought of that scenario."

Jess smiled. "That's because of your optimism Bailey. You think of every problem that hits as a new adventure, not the end of an old one."

Looking at Bailey, Patrick also smiled. "That's what makes you so special Bailey. I wish more people shared in your way of thinking."

Nodding, Joshua sighed. "We'll deal with all that when the time comes. First though, the four of you have a couple trips to make. The chamber of Guardians is planning on making everything public in a couple days. I'd like to be able to add Firmon to the worlds they will be sharing with the people of Terrene."

As the Alpha Team walked to the doorway, Joshua adjusted the controls and then opened the door and watched as the small group stepped into Firmon.

A frown creased Baileys' forehead as she stepped out onto the orange colored soil. Looking ahead at the strange rock formations, Bailey let out a disheartened sigh. "Oh hell, I didn't expect this."

Patrick, who was making his first visit to Firmon, frowned. "What's wrong Bailey? I think this place is spectacular. Look at the size of those boulders. This would be a great place for rock climbing."

Shaking her head, Bailey stared at Patrick. "The last doorway we entered through into Firmon opened into a paradise. It was lush and green. This place is a huge desert."

Patrick laughed. "Some people might consider this a paradise. Everyone has their own ideas for what makes up a delightful world Bailey."

Stepping ahead of the others, Kyle rubbed his hands together. "Let's go take a closer look. I sure wish I had my motorcycle out here. Who knows what's hidden in those rocks? This place is amazing."

Letting out a groan, Bailey stared ahead. "I guess we'll find out."

The group moved toward the rocks, which loomed above them. The tops of some going hundreds of feet in the air. The crew spent two hours exploring the nooks and crannies. After finding a few strange looking creatures whose low bodies and scorpion like features seemed perfectly adapted for Firmons' desert, Bailey sighed sadly. "Let's go back and try the other doorway. This part of Firmon is almost inhabitable."

Kyle shook his head. "Why would you even say that? I think this place is great. It would be better if, like I said, I had my motorbike, or better yet an all-terrain vehicle. I could really get some exploring done then, but for now, I for one, am astonished. I know a lot of people who would feel the same."

Bailey laughed. "I guess Patrick was right. Everyone has a different idea of a perfect world."

Jess slipped a hand through Baileys' arm. "Beauty is in the eye of the beholder. Let's get on back to the lab. We should have plenty of time to make another trip today. If the third doorway in Firmon shows no problems, Joshua can tell the Guardians to put it on their list of alternative worlds available."

With mutual agreement, the four made their way to the doorway and then using their EDAR stepped through into the lab where Joshua waited. He smiled broadly as the Alpha Team stepped in. "What did you think?"

Kyle was nodding enthusiastically, while beside him, Bailey was shaking her head. Joshua laughed out loud. "I take it the two of you have differing opinions."

Stepping over to Joshua, Bailey sank into a chair. "I wasn't impressed, but Kyle was ecstatic. I guess some people could live in that part of Firmon. Personally, I'll take trees and lakes."

Joshua nodded. "If you all feel like it, there's time to check out the other doorway. I think the third area of Firmon will be a surprise to you also."

Bailey wasn't as excited as the others. For her, the first trip to Firmon was the one she wanted to focus on and the place she would love to start a new world.

Standing at the doorway, Bailey closed her eyes, almost afraid to look. When the opening materialized, she felt Jess, standing beside her, nudge her in the ribs. "Open your eyes Bailey, this area is definitely not a desert."

As her eyes opened, Bailey smiled. The landscape in front of her was about as different from the last one as you could get. Stepping forward, Bailey and the others found themselves knee deep in moist vegetation. The trees, they had just stepped under, had broad dark green leaves that hung heavy over their heads.

Kyle wiped his forehead, already damp with perspiration. "Wow, feel that humidity. A little different than that dry desert we just left."

He turned to Bailey. "Is this more to your liking?"

Reaching her hands up, Bailey lifted the hair off the back of her neck where it had already began sticking to her skin. "It's better than the desert, but I'd have a hard time living in this climate. Let's get the exploring over with before I'm nothing more than a puddle on the ground."

The others laughed at the disgust in Baileys' voice. Even though they were all wiping sweat from their foreheads.

As they began walking, Jess shook his head. "I can't believe Joshua didn't give us more warning about this place. We can't stay here long without water to replace what we're losing."

Kyle was the first to move forward into the jungle like area, pushing aside the dense vegetation. The other team members staying as close as possible. Patrick was at the end of the line. The others stopped walking when they heard him cry out. Turning to him, it didn't take long for the rest to see why he'd made the noise.

A large, black snake had wrapped itself around Patricks' arm. By the way his flesh was bulging out from beneath the sides of the snakes' clutches, the thing was trying to kill him with a squeezing motion. Kyle stepped forward, pulling out his stunner and jabbing it against the snake.

Releasing its' hold on Patrick, the creature fell to the ground, motionless. Bailey ran to Patricks' side. "Let me look at your arm. Did that thing break the skin?"

Rubbing his arm as he flexed it, Patrick shook his head. "It didn't break the skin, but for a minute I thought my arm was going to explode under the pressure."
Patrick shuddered and looked at Kyle.
"Thanks Kyle, for a second there…"
Patrick left the sentence unfinished and continued shaking his head.

Looking at the snake, Kyle frowned. "It's a damn good thing that thing didn't wrap itself around your neck instead of your arm."

Jess stepped over and grabbed the snake by its' tail and then threw the six foot long body off into the bushes. "Let's get out of here. I think we've seen enough. I'm not that big on living in a jungle anyway. At least we've found out that some parts of Firmon are inhabitable."

Jess moved closer to Patrick. "We need to take you back and get that arm looked at."

Looking at the welts that were starting to rise where the snake had wrapped its' body around his arm, Patrick nodded. "Yeah, I guess that couldn't hurt."

The crew walked quickly back to where they knew the doorway would be. Jess pulled out his EDAR and activated the entry way.

As they four stepped through, they glanced down, scanning the ground for more creatures. None of them felt good until the door shut behind them.

As they stepped in the lab, Joshua saw the looks on their faces and knew something was wrong. "What the hell happened? You weren't out there for very long."

As Patrick began to hold up his arm, Joshua gasped. "Oh hell, let's get you over to the infirmary."
Turning from Patrick, Joshua looked at Bailey. "Why don't you come with me?" Then he turned to Kyle and Jess. "You two go ahead and log in your information from Firmon. When you're done, shut down the lab and you can meet us at the infirmary. By then, we should have Patrick fixed up."

As Joshua and Bailey took Patrick out of the lab and headed to the infirmary, Bailey filled Joshua in on the trip the team had made to Firmon. By the time she finished, they had made their way through the halls to the faciltys' sick bay.

Bailey was surprised to recognize the woman behind the desk, who stood when she saw the three, her green eyes filled with shock. "Patrick, what happened, are you okay?"

Then Lucy shook her head. "Let me get the nurse before you explain. I'm just filling in."

Lucy hurried away to grab the nurse on duty. Bailey recognized the woman who returned with Lucy. "Hi Megan, looks like I brought you another casualty. This is Patrick."

Stepping forward, Megan looked at Patrick, cradling his injured arm. "I take it you're the one Bailey's talking about."

Patrick nodded. "I had a little run in with a snake. The damn thing hugged me just a little tighter than I wanted."

Joshua looked at Megan. "I'd like to get some x-rays and maybe we can start Patrick on an anti-biotic. I'm afraid we don't know what kind of snake it was that attacked him."

Megan nodded. "Don't worry, I'll take care of everything."
Turning to Lucy, Megan gave her a reassuring smile. "Could you go ahead and check room two for me and make sure it's

ready for a patient? I'll take Patrick down and get some pictures, then bring him there when we finish."

Nodding, Lucy then turned to Patrick. "I'll be waiting for you in your room."

About to nod, Patrick frowned instead. "Where's Miya at?"

Lucy smiled. "Right now, she's down the hall practicing first aid on a doll. It seems while I am doing my training here, I won't have to worry about babysitters. From what I can tell, everyone seems to want to spend time with Miya."

As a grin crossed Patricks' face, he started to shake his head, when a wave of dizziness hit him and he moaned. Beside him, Megan grabbed a hold is arm. "Are you okay?"

Patrick nodded. "Just feeling a little weak. I'm okay, let's go get those x-rays."

As Megan took Patrick down the hall, Bailey turned to Joshua. "I think I'll go with Lucy and wait in the room."

Joshua nodded. "I'll stay here and watch for Jess and Kyle. We'll head in after they get here."

As Bailey followed Lucy to room two, she could see the woman was upset. Bailey knew Patrick had been spending a lot of time with Lucy and Miya. "Patrick's going to be okay Lucy. The people here know what they are doing."

Lucy nodded. "I know, it's just such a shock. This is really only my first day helping out and I was just surprised to see Patrick standing there."

Bailey nodded. "And the two of you have gotten pretty close too, right?"

A shy smile covered Lucy's face. "Yeah, Patrick is wonderful, Miya just loves him like crazy."

Bailey smiled. "What's not to love?"

As Lucy nervously remade a bed that was already neat as a pin, Bailey did her best to explain what had happened. As she finished, Joshua stepped into the room, followed close behind by Jess and Kyle. Looking at the empty bed and then over to the two chairs in the room, Kyle frowned. "Why don't I go find us some more chairs while we wait on the patient?"

When Kyle returned carrying three folding chairs, Jess took two from him and handed one of those to Joshua. The woman took the two seats that had already been in the room.

A few moments later, Patrick and Megan came into the room. She helped a pale Patrick onto the bed. "You just rest, I'm going to go and get you an antibiotic."

When Megan left, Lucy went to the sink and wet down a cloth. Bringing it over, she sat on the bed next to Patrick and gently wiped the perspiration from his forehead. The green eyes were anxious. "You're running a fever Patrick. I think that snake you ran into must have been poisonous."

Joshua was nodding his agreement. "Lucy's right, I think we need to get that antibiotic started and then it might be best to move you to the healing chamber."

Joshua stood. "I'll let Megan know. I think she can give you an intravenous antibiotic. That should get things moving quicker and I think getting you in that chamber will make a world of difference."

Trying to nod his agreement, Patrick found the motion made him dizzy. Instead he half closed his eyes and a hint of a smile touched his lips. Joshua, not liking what he was seeing, hurried out of the room.

Lucy reached over and took a hold of Patricks' arm and shook it. "Stay awake Patrick. We're going to get you some help."

The brown eyes opened then and stared at Lucy as Patrick whispered hoarsely. "I'll try."

Megan and Joshua stepped back into the room. Heading to the bed, Megan quickly hooked up an IV. She looked at Patrick as she started the antibiotics running. "This will just take a few minutes, then Joshua can move you. Luckily, none of your bones are broken. But I'm afraid that snake of yours may have still done some damage."

Megan frowned as she looked down at Patricks' arm. The welts he had earlier had now changed into angry looking, deep red boils that were already festering.

Joshua too had noticed the difference in the wounds. He turned to Kyle and Jess. "Would you two go and get the chamber ready? I'll bring Patrick as soon as that IV is finished."

Jess and Kyle stood, glad to be doing something productive, instead of watching their friend and feeling helpless.

Bailey also stood and stepped over to Lucy. "Would you like me to go and get Miya?"

Looking gratefully at Bailey, Lucy nodded. "I would feel better if she was with me."

Glad for something to do, Bailey nodded. "I'll get her and meet everyone down in the healing room."

*　　　*　　　*

Twenty minutes later in the healing room, Jess was being placed in the chamber. Lucy, with Miya on her lap, watched nervously. Before Joshua closed the lid, he looked over at mother and daughter. "He'll be staying in here at least a few hours, would the two of you like to say anything to him?"

Lucy nodded as she stood, carrying Miya, and stepped over to the container. Miya leaned down first, hanging on to Lucy with one arm and kissed Patrick on the cheek. "Hurry and get better."

As Miya drew back, Lucy also bent down to place a kiss on the cheek. "We'll be waiting for you."

Patrick smiled and although feeling weak and tired was able to whisper. "You better be."

Stepping back, Lucy and Miya watched as Joshua closed the lid to the healing chamber. Lucy shakily moved back to the chair and sat down again, holding Miya on her lap. Bailey, remembering her own worries as she had sat in the same chair, with Jess occupying the place Patrick did now, felt her heart ache for the woman and child. Stepping over, she reached her arms around mother and daughter and gave them both a reassuring hug. "Don't worry, the healing chamber can work miracles. Believe me, I know."

Lucy only nodded, afraid if she spoke, she'd break down and cry.

Kyle, standing by the computer, where he'd just activated the healing chamber, looked at the others. "I could use a cup of coffee, why don't we all head to the cafeteria? Like Joshua said, this will take at least a few hours."

As the group filed out of the room, Lucy turned and blew a kiss toward the chamber where Patrick now finally slept.

Chapter 11

Twelve hours later, Patrick had been removed from the chamber and was back in the hospital room, half way sitting up in the bed, Lucy was sitting next to him. Jess and Bailey had taken Miya with them to their apartment for the night.

Patrick reached out and grabbed Lucys' hand. "Thanks for being here Lucy, it means a lot to me."

Lucy smiled. "There isn't anywhere else I'd rather be."
Then a smile lit up her face. "Well, I'd rather we weren't in this room, but there's no one I'd rather be with."

Patrick also smiled. "I'm glad to hear that. I was worried about asking you something, but I think now I can."

Lucy frowned and Patrick gave out a light laugh. "I'd like to ask if you and Miya would join me in Scintilla. Joshua is moving this project there and I want to continue working with him. Will the two of you join me in that world?"

Feeling her heart soar, Lucy nodded and then stood and nudging Patrick to move over, joined him on the narrow hospital bed. Patrick pulled her into a hug and the two fell into some much needed sleep.

*   *   *

At their apartment, Bailey and Jess sat together at their kitchen table. They'd made a bed for Miya on their couch and now spoke in whispers, so they wouldn't wake the worn out child. Bailey was smiling. "I'm so glad Patrick is going to be okay."

Jess nodded. "So am I. You never know with something that comes from another world. Joshua told me he'll be moving the healing chambers to Scintilla."

Then Jess frowned. "You realize when we get to Firmon, we won't have anything like that. We'll be on our own."

Bailey nodded. "I've thought about that. We'll just have to keep everyone away from that jungle area. The rest of that world isn't even close to what the jungle was. I think we'll be okay. We can transfer some of the medications from here."

Jess smiled. "Along with the hundreds of other things we'll need. I have a feeling we're going to make a hell of a lot of trips through the doorway in the next several months."

Bailey grinned. "Maybe I better start making a list."

Standing from the table, Jess stepped over to Bailey and then leaned down and kissed her. "First though, I think we could use some rest."

As she stood, Bailey took Jess' hand and then the two checked on Miya before heading to their bedroom.

*　　*　　*

Three days later, the Alpha team, along with Lucy and Miya, sat in Jess and Baileys' living room. Jess stood and turned on the large monitor and then sat back down as the screen lit up.

The video feed showed a large table, surrounded by the twelve members of the Chamber of Guardians. The people in the room knew the live feed was going out to every house in Terrene.
At the head of the table, one member slowly stood and faced the camera. His wrinkled face was grim beneath the head full of white hair as he began to speak. "We, the members of the Chamber of Guardians, are here tonight to share horrendous news with the people of Terrene."

As the man continued speaking, the small group sat fixated on his words. First, the man, who spoke for all the guardians, explained about the inevitable impact of the asteroid with Terrene. Then slowly and methodically, the citizens were told about the Postern Project and the hopes of moving the populations to the other worlds that had been found to be habitable.

Although the group in Jess and Baileys' apartment already knew what was going to be said, it was still a shock to hear it stated. As they listened, the daunting challenge ahead of them hit home.
As the speech ended, Kyle was the first to shake his hand and voice his opinion. "Looks like our few days of rest and relaxation are over."

Jess nodded. "There's so much to do. Has everyone made a final decision on what world they'll be going to?"

Bailey frowned. "I was hoping our team would all head to the same world."

Turning to Bailey, Patrick shook his head. "I'm not going to Firmon Bailey."

Staring at Patrick, Bailey raised her eyebrows. "Is it because of that snake?"

Patrick laughed. "That would be a good enough reason. No, it's because I want to help Joshua in Scintilla. You know how I am about computers. I don't think I could even live in a world without them."

Looking over at Miya, sitting on the floor, Patrick held out his arms. Standing, Miya crawled happily into Patricks' lap. Smiling, Patrick turned to the others. "Believe it or not, Lucy and Miya have agreed to join me. I can't tell you how lucky I am."

Lucy, sitting next to Patrick, took his arm. "Miya and I are the lucky ones."

Sighing, Bailey nodded. "I'm happy for all of you. Although, I wish you were coming to Firmon."
She turned to Kyle. "You're going to Firmon aren't you?"

Kyle nodded. "I'll be there, but I'm more interested in the desert. In fact, I've already arranged to have a few prototypes of solar powered motorcycles delivered. They're going to be perfect to use in Firmon."

Bailey stared at him. "What a great idea. I think I better take the time to check out what else Terrene has to offer in solar powered items. They'll be perfect for Firmon."

Jess was nodding. "Just don't forget, anything we take in, has to fit through the doorways. Some things will have to be taken through in pieces and then assembled on the other side of the doorways."

Bailey laughed. "That's what I have you and Kyle for."

Kyle rolled his eyes. "I guess I better start recruiting the strongest people I can find to join us in Firmon."

The others laughed, but Lucy was frowning. "I've heard your team say these parallel worlds are connected. What does that mean for the other worlds when that asteroid hits Terrene? Won't it have an effect on those worlds? What if we do get settled in these other places and the destruction happens in those worlds too?"

Patrick put an arm around Lucys' shoulders. "We've run thousands of scenarios through our computers. In all of our simulations, we haven't found the destruction will cross the worlds."

Jess was nodding. "The other worlds may feel some repercussions, but nothing that would make a difference, that is, if it is felt at all."

Bailey shook her head. "Until it happens though, we have no way of knowing with any kind of certainty, what will happen. I'm sure cataclysmic events happen all the time and here in Terrene, we have no clue what is happening in the other worlds."

Patrick smiled. "If all goes well in Scintilla setting up the Postern Project, we can still continue with the traveling, including into Firmon, where these wild ones will be making their home."

Lucy smiled. "I sure hope so, I'd like to be able to visit all of you."

Chapter 12

The next month seemed to fly by, as not only the teams that worked at the Postern Project, but also the people of Terrene began to begin making preparations and started moving into the other worlds.

For those going to any world, but Firmon, the worst part of the move was knowing they had to leave so much behind. When compared to eminent death, if they stayed in Terrene, the decisions were easier for the few traveling to Firmon. For them, the hardest part of their journey was finding everyday items that could be converted to solar or wind power.

Several trips were made through the doorways each day. The first move for Kyle had been to bring a half dozen of the solar powered cycles through the doorway.

When the six special machines had first been supplied to the facility, Bailey had laughed at Kyle. He was more excited about the delivery than Miya had been at the prospect of shopping for new toys when she and Lucy had first come to Terrene.

Bailey and Jess had been in the lab with Kyle the day the bikes had been delivered. Kyle had been called to the facilitys' garage and had asked Bailey and Jess to come with him to check out his new toys, as Bailey called them. When the trio had stepped in the garage, Kyles' eyes had lit up as he stared at the six new machines. To Bailey, they looked like ordinary motorcycles. When she said as much to Kyle, he had shaken his head in disbelief.

"Are you kidding me? These bikes are like nothing you've ever seen before. That don't need a battery that has to be recharged, or fuel to keep running. The sun does all the work. The desert area of Firmon is nothing short of perfect for these babies."

Kyle rubbed his hands together. "I can't wait to take them through the doorway and try them out."

It hadn't taken long for Kyle to convince Jess and four more people that had chosen

Firmon for their new world, to each take a bike through into Firmons' desert area.

Instead of joining them, Bailey had connected with a dozen people and they had traveled through the other doorway to Firmon, where her paradise lay.
Baileys' group carried tents, solar lanterns and boxes of seeds, making trip after trip, as they slowly built up their stockpile. Each day, more and more people were added to the list Bailey had begun keeping. She was using a binder where she wrote down not only the new residents' names, but where she added their birth dates, family members and most importantly, each persons' special skills. She was exuberant as she saw how many of the people that opted for Firmon brought with them farming and gardening backgrounds. Skills like that would come in handy as they built a new world from almost nothing. People who were looking for technology had decided to try Scintilla. They joined Joshua and Patrick as they began the building of a new Postern Project in that world.

By the time two months had passed, since the Chamber of Guardians had shared Terrenes' future with its' population, a third of the people from Terrene had relocated into the six worlds. Joshuas' three teams had discovered and added their numbers to those worlds. The refugees carried with them the bare minimum of necessities and keepsakes as they began their new lives.

It was at that two month mark the Alpha team met with Joshua, Lucy and Miya in the facilities' cafeteria.

As they took their seats at the table, Joshua remained standing. A broad smile covered his face. "First of all, thanks everyone for taking time out of your hard work to be here. Second, thanks for that hard work you've all been doing. I think in the last two months we've accomplished nothing short of a miracle."

Taking his seat, Joshua took the time to look at each person at the table before he continued, his words rushing out with enthusiasm. "I'd like to hear from Jess, Bailey and Kyle about what they've been able to accomplish in Firmon and how people are adapting to their new world. I hope they are doing as well there as we are in Scintilla and as the reports are showing the Beta and Gamma teams are doing in the others."

Jess laughed. "We've accomplished a lot. People are settling in, but why don't you share what's going on with you first, Joshua. We can see you're excited. What have you and Patrick gotten done with your project?"

Nodding, Joshua laughed. "Glad you asked. We've moved the basics needed and have finished the first test on opening a doorway from Scintilla. I think you all know before now, the doorways had to be opened from our lab at the Postern Facility in Terrene or by using the EDAR Units. Yesterday, we actually were able to open the doorway from Scintilla to step through into the lab in Terrene."

Looking at Joshua, Baileys' face scrunched in curiosity. "What about opening a doorway from Scintilla into Firmon. After the asteroid hits, we can no longer reach you. Have you figured out a way to fix that?"

It was Patrick who nodded. "We think we might have come up with the program for that. You've been busy using the doorway from the lab to the area where you and Jess have been setting up, so we thought it would be best to open the doorway that leads to the desert, where I hear Kyle's been busy running around on that solar bike of his."

Kyle was smiling and nodding. "You heard about that? Sounds like I'm famous."

Shaking his head, Patrick also laughed. "Maybe infamous would be a better choice of words. From what I've heard, you've turned into a regular daredevil on that bike, you and your group of riders."

Kyle pointed at Patrick. "Tell you what, you get that doorway opened and I won't even charge you to come and watch me."

Bailey shook her head and turned to Lucy. "Those two are crazy. I'd rather hear about you and Miya. How are you settling in?"

Lucy smiled. "I really like it in Scintilla, but Miya loves the place, right Miya?"

The young girl nodded, her eyes open wide and sparkling. "I get to start school soon, but even better than that, we have a big house. It has a swing set and a slide. There's two girls that live next door." The eyes opened even wider. "And they're my age. The girls are twins, they look exactly the same."

Bailey couldn't help but laugh at Miyas' enthusiasm. "Sounds to me like Scintilla is the perfect place for you. I'm so happy it's working out so well."

As Lucy nodded her agreement, she turned and exchanged a glance with Patrick, who grinned and nodded his head.
Turning back to Bailey, Lucys' eyes, so like her daughters seemed to glow with an inner light. "We have even more good news. Patrick and I are going to get married."
Looking at the couple, Bailey realized now what the exchanged glance had meant.
Then Patrick cleared his throat. "We have one more surprise to share."

Before Patrick could explain, Miya bounced up and down on her seat. "Can I tell, can I?"

As both Patrick and Lucy nodded, Miya exclaimed with delight. "I'm going to be a big sister."

Kyle laughed and shook his head. "You're a fast worker Patrick, congratulations, to all of you."

Turning to Lucy, Bailey smiled. "That is great news. I'm so happy for all of you."

Taking Baileys' hand, Lucy half shrugged. "If it wasn't for you, who knows what might have happened to me or Lucy? You saved our lives back in Province."

Joshua interrupted on the congratulations going on. "I'm just glad we all found the spare time today to get together and share what's happening in our lives. I'm afraid though, I really need to get back to work." He looked at Patrick.
"Sorry, but I really am going to have to take you away from your friends, I need your help to prepare for the testing on the

program to open the doorway into Firmons' desert."

Joshua turned to Kyle. "I want to give it a trial run tomorrow afternoon."

Kyle nodded. "I'll be on my bike and waiting."

As Patrick and Joshua stood, Lucy took Miyas' hand. "C'mon honey, we're going too."

The small lips puffed out in a pout before Miya stood and ran over to Bailey to give her a hug. When she finished, Miya turned and hugged the others at the table.

As the rest of the group stood and following Miyas' example exchanged hugs and handshakes, Bailey smiled and wiped away a tear. Then she shook her head at her emotions. She didn't know where that had come from. It wasn't like she wouldn't be seeing her friends again. With Joshua making the breakthroughs he was in Scintilla, she'd be seeing all these people soon, not saying goodbye forever.

After half the group that had been seated in the room left, the other half retook their seats. Looking at Jess and Kyle, Bailey sighed. "I can't believe how hard it was to say goodbye."

Jess laughed. "It won't be for long."

Kyle nodded. "Jess is right, I can't wait to have Patrick and Joshua come through that doorway and see what I've been doing out in the desert."

Pushing back from the table, Jess got to his feet. "I think it's time for us to get going to. We have a lot of work to do in Firmon. I know we left others in charge, but I always feel better when I'm there watching what's going on."

Kyle nodded. "So do I, besides, I want to set up something really spectacular for when Patrick and Joshua come through that doorway."
Heading back to the lab, Bailey walked between to two men. "I can't believe how much has been accomplished. I really think

we are going to be able to save the people of Terrene."

Jess smiled. "We still have a long way to go and a lot to do, but I'd say at the rate things are going, we'll have Terrene almost empty ahead of schedule. I'm just glad we have the Beta and Gamma teams to help. If you think about it, they've moved more people than our team has. We've been able to put quite a few people in Scintilla, but compared to that number, Firmon is way behind."

Bailey smiled. "We knew from the beginning Firmon wasn't a place many would choose. I'll bet by the time we're done though, Firmon will be able to count at least a thousand people who want to make their home there."

Jess shrugged. "Maybe, I'm actually surprised by the number of people who have already picked Firmon over the other worlds."

The trio had reached the lab door. Kyle pulled the door open to let Bailey and Jess step in ahead of him. As he closed the door,

Kyle pulled out his EDAR Unit and moved to the doorway where Jess and Bailey now stood. Holding the unit in front of him, Kyle smiled. "I'm just glad Joshua got these units to us. It sure makes the traveling a lot easier."

Jess laughed. "I wish I would have had one that day in the Earth World. I can remember seeing that distortion in the air from the ground. I was so damn close to home and had no way to open that doorway and step through."

A shiver ran through Bailey, thinking how close she'd come to losing Jess. She felt her partners' arm slip around her. "Sorry Bailey, I didn't mean to dredge up bad memories."

Looking up at him, Bailey nodded and smiled. "It's okay, that's past history and can stay there. I think the future looks much brighter and more worthy of my focus." With that optimistic thought, the three stepped through the doorway.

Chapter 13

The next day, Bailey and Jess had already taken a group of twenty people through the doorway. They'd already gone to the same town Bailey had taken Lucy and Miya when they'd first ended up in Terrene.
The group had gone to a large warehouse that stockpiled disaster preparedness supplies. Stores, like the one the group visited, had begun opening in Terrene when, years ago, a flood had disabled one area of their world. The relief kits the stores carried contained items meant for short term survival for smaller disasters. No one then, except maybe Joshua, had envisioned the enormous disaster that loomed ahead of all them in the future.

After loading what had once been an appliance delivery truck with supplies, the group returned to the Postern Facility and transferred the supplies onto work dollies. Then the procession began as they wheeled the much needed supplies into the lab and through the doorway into Firmon.

For now, everything was stored in various tents that had been set up. The tents, most the size of a small apartment, dotted the landscape.

Closer to the lake, more tents, used as temporary living quarters, were occupied by the new, slowly growing, population of Firmon.

In the tent, shared by Bailey and Jess, the two were having a quick lunch while they took a break between loads.
The couple sat on chairs that could easily be folded and had been brought through two months ago when all this had begun.

Jess was staring at Bailey across the small table that also could be folded and moved out of the way. "You really look beautiful, Bailey."

Running a hand through her brunette hair, Bailey shook her head. "You're crazy, I must look a mess. After all the work this morning, I could probably use a good wash too."

Shaking his head, a smile lifted Jess' lips. "I don't know about that, but I think setting up a new world agrees with you."

Laughing at that, Bailey then shoved the last of her sandwich in her mouth and stood. "Which reminds me, we better get back to the job or it will never get done."

Jess stood next to her. "We're making good progress, but where I'd really like to be is out in that desert with Kyle. Getting a doorway open between here and Scintilla is going to be a major breakthrough. Once that asteroid hits, the doorway to Terrene will be closed forever."

Bailey moaned at that thought. They'd gotten used to being able to travel to civilization in just a couple short steps. Soon, that avenue wouldn't be found. "I can't begin to imagine the kind of show Kyle has in mind for Patrick and Joshua."

Jess laughed. "Knowing Kyle, it's going to be something spectacular. Maybe he'll ride one of those solar bikes up the side of a rock mountain."

As Bailey pictured Kyle doing just that, she nodded. "That's probably just one of many tricks Kyle will try. I'm sure the desert today will have a show to remember."

*   *   *

In the desert, Kyle was busy trying to set up what Jess and Bailey had been talking about. He wasn't sure exactly when Patrick and Joshua would be opening the doorway, but he wanted to be ready. Kyle had three men and one woman practicing tricks with him on the solar bikes.

Between their own trips from Firmon to Terrene to bring supplies, the five of them had spent every few moment on the bikes. Until now, that had been just because they shared the love of riding the bikes.

After talking to Joshua and Patrick the day before, Kyle had returned to the desert in Firmon and asked the others to help him prepare what he hoped would be a glorious show.

Instead of taking time off the bikes for lunch. Kyle and the other riders had only stopped long enough for a sandwich. That had been scarfed down while the group stayed sitting on the bikes.

Looking up at the sun, Kyle knew it was about time for the doorway to be opened. As the others continued to practice, Kyle rode his bike closer to the area where the doorway would be opened.

Having ridden motorbikes for years, Kyle loved the innovative solar bikes. Nothing compared to the sleek and silent bike he'd brought through into Firmon. The bikes were also lightweight, without the heavy engine of a normal motorbike. In fact, that, to Kyle, was the best part of the design. The bikes were not only faster, but doing a trick, like a three sixty flip, was so easy, it was almost shameful.

Kyle smiled at the thought and then stared at the spot where he could see the barely discernible outline of the doorway.
He waited anxiously for it to open.

Kyle had wanted Jess and Bailey to be here. He knew they would have been, but they hadn't wanted to use the desert doorway to get here. Unsure of when Joshua and Patrick would need it for their own traveling through.

Even with the bike Kyle had left for them to use, it would have taken them four hours to reach the desert. Using the doorways and the ability of instantaneous travel had spoiled all the teams and made normal travel seem even longer.

Staring impatiently at the area, Kyle felt the ground shake beneath him. The tremor or whatever was causing the ground to move, only lasted a few moments, but had surprised Kyle in its' intensity.

As the shaking stopped, Kyle let out a sigh of relief as the doorway in front of him finally opened.

He could see through the opening and smiled at the sight of Joshua standing in the entrance. Kyle could just make out the enough of the background to see Joshua was in the computer room in the basement of the Serenity Center in Scintilla.

As Kyle saw the distraught look on Joshuas' face, his smile vanished.

*   *   *

Four hours away, Jess and Bailey were unloading boxes into a tent, when they also felt the movement of the land beneath their feet. Frowning, Bailey then stared wide eyed at Jess. They could hear a few astonished cries from others outside the tent.

Stepping outside the tent, Bailey and Jess moved out to join the others. By then, the slight shaking had stopped. Jess looked around, seeing the frightened looks on the faces around him. He held up his hands. "Nothing to worry about. Just a little tremor. It's over now. This area was researched thoroughly. Although we did find spots where we could see some fault lines, none of those were even shown close to this area. Back in Terrene, we've all seen much worse. That was nothing more than a little hiccup. I promise you, we're safe here. The best thing we can all do right now is just get back to our work."

Slowly, the others, reassured by Jess' words, went back to their job of unloading supplies.

Jess and Bailey went back into the tent. Seeing the worried look on Baileys' face, Jess pulled her into a hug. "Don't worry, I was telling the truth out there. We're safe here."

Bailey nodded and then stepped back from Jess, still frowning. "I was thinking about Kyle and the others out in the desert. Do you think they felt the tremor out there? Some of those rock formations looked pretty unstable to me."

Frowning, Jess finally tried to give Bailey a smile. "I'm sure they're fine. Kyle knows how to take care of himself."

Although Bailey nodded, she wasn't as convinced as she'd like to be. "I wish we would have gone out to the desert today."

Shrugging, Jess shook his head. "It's at least a four hour ride Bailey, and that's just one way. You know how much we have left to do. It just wouldn't seem right to take that much time off our work. I'm sure Kyle will show up this evening. He'll come riding up after Joshua and Patrick get done over there. Until then, the best thing for us to do is get back to work."

Bailey nodded, glad for something to keep her mind occupied and maybe stop her from worrying.

*   *   *

By the time Jess and Bailey sat down for their evening meal, the unloading had been finished for the day. Both also knew they had a lot more days of the same procedure ahead of them. Jess stepped outside to grab two beers from a solar powered cooler.

Stepping back into the tent, Jess handed one to Bailey, before he opened the other and took a long drink. Then he took the seat across from her at the small table.

Bailey was only picking at her dinner. Knowing what was bothering her, Jess tried to give her a comforting smile. "I'm sure Kyle's okay. The others in the desert with him too. Listen, if he doesn't show up tonight, I'll load you on that solar bike and take you to the desert in the morning. Checking on Kyle is a lot more important than how much stuff we can get moved. Besides, I know until we check on him, you won't be able to concentrate on the work here anyway."

The blue eyes brightened. "Really?" Then she shook her head. "Sorry, I just thought Kyle would be here by now and bragging about how much he impressed Joshua and Patrick with his riding skills."

Jess laughed. "I can just imagine Kyle doing that too. Now, if you're done picking at your food, why don't we take a walk?"

Pushing away the plate she hadn't touched, Bailey took a drink of her beer. "That sounds like a great idea."

Carrying their drinks with them, the two stepped out of the tent. As they walked, the couple smiled and waved at several people who were sitting outside, resting from their own long day of work.

Heading away from the tents people were using as temporary homes, Jess and Bailey made their way toward the doorway to Terrene. The tents holding supplies instead of people had been set up in that area. Looking ahead of them, Bailey let out a squeal of excitement and pointed. "Look over there Jess, isn't that Kyle?"

Seeing the lone figure on a solar bike, Jess nodded. "It is, I'm sure he's come to brag about his day. He sure took his sweet time getting here."

Walking toward Kyle, the two stopped in their tracks when he got close enough for them to see his face covered in anguish.

The beer Bailey had been carrying slipped unnoticed from her hand.

Screeching to a halt, Kyle jumped off the bike, letting it drop to the ground. He ran to the couple. "It's gone, oh hell, Terrene is gone."

Seeing the frowns on his friends' faces, Kyle drew in a deep breath. "Joshua came through the doorway to tell me. The damn asteroid hit." Kyle wiped a hand over his face. "They're dead. Everyone left on Terrene is dead."
The brown eyes closed a moment, when they reopened, they were filled with tears. "Patrick was in Terrene, he had Lucy and Miya with him."

Staring at Kyle, Bailey felt her stomach do a flip flop. She dropped to her knees, afraid she was going to throw up. Jess knelt beside her, rubbing her back, but looking up at Kyle. "What the hell are you talking about? We have months until the asteroid hits. You must have heard wrong."

Kyle shook his head. "Joshua saw it Jess, he was looking through the viewer in Scintilla. He'd just sent a group to Terrene. Patrick, Lucy and Miya were in the group. The doorway closed and Joshua watched the group through the viewer step into the lab on the other side of the doorway and then... and then..."

Feeling like he might be sick himself, Kyle drew in a couple of deep breaths.

"Then Joshuas' screen showed an explosion and went dead. He tried to make contact, but there's nothing or no one left."

Kyle pulled his EDAR Unit from his pocket and held it up. "This unit is dead. It was linked to the Postern Facility. When Terrene blew, the lab disappeared and so did our link to that place. Thankfully Joshua had the link from Scintilla to here in his computer, or we would have never known what was happening."

Reaching in his own back pocket, Jess pulled out his own EDAR. Trying to turn the unit on, but seeing only the blank screen, he threw it to the ground in disgust and despair. "Why didn't we see it coming? All our research, we had the impact down to almost the exact minute. What the hell happened? Kyle, could Joshua be wrong?"

Shaking his head, Kyle ran his hands through his hair. "He's not wrong Jess. The computers were wrong and now that world and the people are gone."

Beside Jess, Baileys' body was jerking with each jagged sob that racked her body. "No, no, this can't be happening."

As Kyle dropped to his knees, Jess pulled Bailey into a hug. Kyle stared at Jess over the top of Baileys' head. "I'm sorry I had to be the one to tell you this. Joshua's still back in Scintilla. He's still searching for any information he can find, but with the lab and computers gone…"
Kyle shook his head. "…there's nothing to find."

Looking at Kyle, Jess' own eyes filled with tears. "I can't believe it, I just can't believe this has happened."
Jess frowned. "That trembling we felt, that

had to be from the impact, oh hell, it really is gone."

Bailey turned away from the two men and now she did throw up on the grass as she remembered that Lucy had also been pregnant. When nothing was left to throw up, Bailey wiped trembling fingers across her lips. "Patrick and his new family are together forever now."
Tears rolled down Baileys' face. "At least they were together. They won't have to live without each other."

As Jess stood, he pulled Bailey up with him. Kyle also stood and stared toward the encampment. "We're going to have to tell them what happened. They're entitled to the truth. Most of them still have family and friends who were back on Terrene."

Bailey leaned against Jess for support as the three friends made their way past the supply tents and then over to where the new residents of Firmon had set up their temporary homes.

Jess enlisted the help of a few people, asking them to spread the word that he needed the people to gather for some important news he had to share.

Slowly, groups of people congregated near where Jess, Bailey and Kyle were standing. Luckily for Jess, only about two hundred people had relocated to this section of Firmon. As they waited for the crowd, Kyle told Jess that Joshua was going to be sharing the same news with the almost fifty residents of the desert. The same sad news that Jess was preparing to share.

Jess waited another ten minutes, when he didn't see any more people coming toward the area, he stepped up on a large rock and cleared his throat. "I'm afraid I have some devastating news to share and unfortunately I don't know an easy way to tell you. The asteroid we all knew was headed for Terrene arrived earlier than we all expected. At the time all of us felt the earth shake earlier today, the world of Terrene was hit and destroyed by that asteroid."

Jess swallowed hard before continuing. "I regretfully have to tell you, no one survived the destruction and what was once the world of Terrene is now gone. I'm so sorry to any of you who had friends and family there."

The three remaining members of the Alpha Team could hear the murmuring of voices, intermingled with several cries of pain and agony. Jess shared their pain. "I think right now, the best thing we can do, is stay here and mourn together the precious lives that were lost."

Slipping down from the rock, Jess joined Bailey and Kyle. Then the three made their way through the crowd trying to give comfort to as many as they could and accepting comfort in return.

Chapter 14

A week later, Joshua traveled to the desert from the doorway he had opened in Scintilla. Borrowing one of the solar bikes, he rode next to Kyle, who was on his own bike. The two men made their way to the encampment.

When they got there, they found Jess and Bailey down by the lake. This was only the second time the couple had seen Joshua since the asteroid had hit.

Bailey gave both Kyle and Joshua big hugs before the two men followed her and Jess into their makeshift home.

Jess unfolded a couple of extra chairs for Kyle and Joshua to use, then he and Bailey also took seats.

His green eyes full of concern, Joshua stared at Bailey. "How are you holding up Bailey?"

Lifting her shoulders, Bailey shrugged and shook her head. "I miss everyone so much. Besides, Patrick, Lucy and Miya, we all had family and friends who were still in Terrene, We are all trying to stay busy here, which seems to help a little. I think it will be a long time though, before the hurt even begins to ease up. How about you Joshua?"

The green eyes that stared at Bailey, darkened slightly. "Like you, I'm taking it one day at a time. I still can't help but feel guilty about the time line for the impact. I still can't understand how our computer models could have been so far off. Those deaths are my burden and always will be."

Jess stared at Joshua, "No way Joshua, you can't take the blame for that, we all looked at those computer simulations. I don't know what happened to move that asteroid closer so fast, but it wasn't anything you had any kind of control over."
Jess frowned. "Have you been able to contact any of the other worlds yet?"

The head of sandy hair shook back and forth slowly. "Not yet, but I will. I have to let the people who left Terrene know what happened. I owe them that at the very least. I miscalculated when that asteroid would hit and now over a hundred thousand people have lost their lives."

Bailey stared at Joshua. "Jess is right, you can't blame yourself. The computer models were wrong and that's the information you were going by. Without your project, everyone in Terrene would have lost their lives, you should remember how many people you were able to save."

As Joshua stared at Bailey, he sighed. "I didn't save enough, not nearly enough." Then he shook his head. "I'm not giving up though. I will find the four worlds where the people of Terrene traveled to and let them know what happened. Right now, the only door we have opened is the one to your desert. Actually, the doorways are the reason I'm here. I don't know what happened to the Beta and Gamma teams, but I wondered if you three would like to resurrect your Alpha team and help me."

Bailey and Jess turned to each other and exchanged glances before turning back to Joshua and shaking their heads. It was Jess who answered for both of them. "No more traveling for us. We've decided to settle here and start a family."

Understanding their need for that, Joshua turned to Kyle, who smiled. "If you can promise me time off once in a while to come back and ride around in the desert, I'd be glad to come and work with you. What about after you find those four worlds where those from Terrene went to start their new lives, will you continue to travel?"

Joshua nodded. "I hope to, you never know when we'll find a world in trouble. And maybe, just maybe, some day, I'll find the one special doorway that leads to the world where all those who not only died in that horrific catastrophe, but others who passed away before, have ended up. I know in my heart that world is out there somewhere. A place where loved ones who have died now live."

As Kyle and Jess nodded, Bailey reached her hand forward to touch Joshuas'. "I know that world is out there too. I also know if anyone can find it, you can."

Bailey smiled. "Especially with Kyles' help. When you do find it, I hope you will come back here and share what you find. I know I have a lot of people I would love to see again."

A broad smile covered Joshuas' face. "You can bet on it."

Thanks readers for grabbing this book and joining me on this journey.
Although the book is fiction and yes, science fiction, I hope someday a trip to another world is just a step away as we go into those doorways.
I also hope the doorways you enter, both now and then, will bring you what your heart desires.

As always thanks to family for their never ending support and thanks to friends, near and far, on-line and off, who have helped me promote and share the stories that seem to never stop in this crazy mind of mine.

Here's to many more journeys.

P.S. Winn

Other titles from P.S. Winn

Novels

Foretold
Voices
Obligations
Tunnels
Capernicious
B.A. 47
Pacific Passage
Suppression
Lies in Shadows
Phases
Mystic Valley
The New Moon Killer
Healings
Superstition Canyon
Collisions
Viewings
At Hidden Lake
Parallel Adventures - Into the Caves
Parallel Adventures - Secrets Revealed
A Gradual Decline
Judgments
Of Jeebies and Journey

Collections

Visitations
Heartfelts
Stretched Stories
Stretched Stories 2

Comic books

The Golden Years
The Golden Years 2

For Children

The Alphabet Book
The Number Book
The Secret Life of Goats
No, Jimmy, No